Poetics of Deconstruction

Also Available from Bloomsbury

Radical Animism, Jemma Deer
Capitalism's Holocaust of Animals, Katerina Kolozova
The History of Animals, Oxana Timofeeva
Philosophical Posthumanism, Francesca Ferrando
Deleuze and the Schizoanalysis of Feminism,
ed. by Janae Sholtz and Cheri Lynne Carr

Poetics of Deconstruction

On the Threshold of Differences

Lynn Turner

BLOOMSBURY ACADEMIC
LONDON • NEW YORK • OXFORD • NEW DELHI • SYDNEY

BLOOMSBURY ACADEMIC
Bloomsbury Publishing Plc
50 Bedford Square, London, WC1B 3DP, UK
1385 Broadway, New York, NY 10018, USA
29 Earlsfort Terrace, Dublin 2, Ireland

BLOOMSBURY, BLOOMSBURY ACADEMIC and the Diana logo are
trademarks of Bloomsbury Publishing Plc

First published in Great Britain 2021
This paperback edition published in 2022

Copyright © Lynn Turner, 2021

Lynn Turner has asserted her right under the Copyright, Designs and Patents Act,
1988, to be identified as Author of this work.

For legal purposes the Acknowledgements on p. ix constitute an
extension of this copyright page.

Cover design by Charlotte Daniels
Cover image: White God (2014) directed by Kornél Mundruczó
(© Magnolia Pictures)

All rights reserved. No part of this publication may be reproduced or transmitted in
any form or by any means, electronic or mechanical, including photocopying,
recording, or any information storage or retrieval system, without prior
permission in writing from the publishers.

Bloomsbury Publishing Plc does not have any control over, or responsibility for,
any third-party websites referred to or in this book. All internet addresses given in
this book were correct at the time of going to press. The author and publisher regret
any inconvenience caused if addresses have changed or sites have ceased to exist,
but can accept no responsibility for any such changes.

A catalogue record for this book is available from the British Library.

Library of Congress Cataloging-in-Publication Data

Names: Turner, Lynn, 1968- author.
Title: Poetics of deconstruction: on the threshold of differences/Lynn Turner.
Description: London; New York: Bloomsbury Academic, 2021. |
Includes bibliographical references and index.
Identifiers: LCCN 2020026419 (print) | LCCN 2020026420 (ebook) |
ISBN 9781350128590 (hardback) | ISBN 9781350128606 (ebook) |
ISBN 9781350128613 (epub)
Subjects: LCSH: Art–Philosophy. | Poetics. |
Deconstruction. | Feminist theory.
Classification: LCC N71.T87 2001 (print) | LCC N71 (ebook) | DDC 701–dc23
LC record available at https://lccn.loc.gov/2020026419
LC ebook record available at https://lccn.loc.gov/2020026420

ISBN: HB: 978-1-3501-2859-0
PB: 978-1-3501-8553-1
ePDF: 978-1-3501-2860-6
eBook: 978-1-3501-2861-3

Typeset by Integra Software Services Pvt. Ltd.

To find out more about our authors and books visit www.bloomsbury.com
and sign up for our newsletters.

To you

Contents

List of Figures	viii
Acknowledgements	ix

1	In the beginnings: Introducing *Poetics of Deconstruction*	1
2	The Animal Cure: Inhaling the other in *Dean Spanley*	7
3	Raising animals: Between the basement and the kennel in *The Woman*	27
4	Speculations: Gesture in *Conceiving Ada* and *Absent Presence*	53
5	'*Unfamiliar Unconscious*': The performativity of *Infinity Kisses*	81
6	Outlaws: Towards a posthumanist feminine after *Dancer in the Dark*	117
7	In lieu of conclusion: The cardio-pedagogy of *White God*	137

Notes	164
Index	208

List of Figures

2.1	Inhale (*My Talks with Dean Spanley*, 2008, Dir. Toa Fraser, UK/New Zealand: Icon Films)	14
2.2	Exhale (*My Talks with Dean Spanley*)	21
2.3	Cry (*My Talks with Dean Spanley*)	22
3.1	Bite (*The Woman*, 2013, Dir. Lucky Mckee, USA: Moderncinéma)	27
3.2	Introduction (*The Woman*)	34
3.3	Pet (*The Woman*)	50
4.1	Bird (*Conceiving Ada*, 1997, Dir. Lynn Hershman-Leeson, USA: Hotwire Productions)	70
4.2	Devil Angel (*Conceiving Ada*)	71
4.3	Braids (*Absent Presence*, 2005, Dir. Hussein Chalayan, UK/Turkey: BM Contemporary Art Centre)	75
4.4	Foreigners (*Absent Presence*)	78
4.5	Splashback (*Absent Presence*)	79
5.1	Carolee Schneemann *Infinity Kisses* I, 1981–87. Courtesy of the Estate of Carolee Schneemann, Galerie Lelong & Co., Hales Gallery, and P•P•O•W, New York © Carolee Schneemann	87
5.2	Carolee Schneemann *Infinity Kisses* I, detail	88
5.3	Carolee Schneemann *Infinity Kisses II*, 1990–98, 2004 print edition as shown in *Remains to Be Seen*, CEPA Gallery, Buffalo, NY, 2007. Courtesy Galerie Lallouz	115
6.1	Factory (*Dancer in the Dark,* 2000, Dir. Lars von Trier, Denmark: Zentropa Entertainments)	126
6.2	Hear (*Dancer in the Dark*)	128
7.1	Courtyard (*White God*, 2014, Dir. Kornél Mundruczó, Hungary: Proton Cinema)	140
7.2	Toast (*White Dog*, 1982, Dir. Sam Fuller, USA: Paramount Pictures)	143
7.3	Blood (*White God*)	152
7.4	End (*White God*)	162

Acknowledgements

The origins of these *Poetics* are, naturally, wildly overdetermined: my parents have taught me more than I realized; sitting in the company of cats has been more conducive to the mood of writing than I would ever have guessed. Barely a trace remains of my doctoral days, many moons ago, living on within this 'first book', yet I hope that Barbara Engh will enjoy these new constellations (while 'that fox' does not make any appearance, many other animals have now seized the day) and that Adrian Rifkin will still appreciate my 'illegitimate use of figure'.

Huge thanks go to Liza Thompson and Lisa Goodrum at Bloomsbury for their enthusiastic shepherding of this book. Many elements of it were honed through presentations in various contexts (thanks to: Lisa Blackman, Nicholas Chare, Laura Cull, Carolin Eirich, Katve-Kaisa Konturri, Olga Koroleva, Pat Naldi, Tutta Palin, Henry Rogers, Undine Sellbach, Ika Willis, Joanna Zylinska). Eszter Timár deserves special mention since we encountered an array of rollicking hedgehogs on the tour of Budapest with which she welcomed me. Four chapters have been significantly revised and extended from previous incarnations. Chapter 2 first appeared as 'Animal Melancholia: On the Scent of *Dean Spanley*', in *Animal Life & the Moving Image*, Laura McMahon & Michael Lawrence, eds. Palgrave/BFI Publishing (2015): 134–52. Chapter 4 first appeared as 'UnHoming Pigeons: The Postal Principle in Lynn Hershman-Leeson & Hussein Chalayan' in *Derrida Today*, 5.1 (2012): 92–110. Chapter 5 first appeared as 'When Species Kiss: Some Recent Correspondence between *Animots*' in *Humanimalia: A Journal of Human/Animal Interface Studies*, 2:1 (2010): 60–85. Chapter 6 first appeared as 'Tympan Alley: Posthumanist Performatives in *Dancer in the Dark*', in *Derrida Today*, 6:2 (2013): 222–39. Chapters 4 and 5 benefitted from the support of the AHRC (Research Leave Scheme, 2007–8), responsible for the 'research surprise' that decisively reoriented my work.

Those who have published my work over the last ten years have offered invaluable advice and also been extremely welcome figures of support (Nicole

Anderson, Ivan Callus, Istvan Csicsery-Ronay, Stefan Herbrechter, Tora Holmberg, Michael Lawrence, Laura McMahon, Simon Wortham, Manuela Rossini, Julian Wolfreys). To those who listened, heard, read, recognized, remarked, thank you: Ron Broglio, Lindsay Kelley, Vicki Kirby, Ted Geier, Johnny Golding, Helena Hunter, Elissa Marder, Dawne McCance, Jean Paul Martinon, Kelly Oliver, Silke Panse, Wood Roberdeau, Nicholas Royle, Astrid Schrader, Susan Schuppli, Undine Sellbach, Elina Staikou, Kari Weil, Sarah Wood, Cary Wolfe.

Of my former students, I especially thank Ruth Lipschitz, whose now postdoctoral thought continues to be instructive. Lastly, to those taking my 'Sex Gender Species' module at Goldsmiths, holding a space in which to allow theory to air the ludic and ludicrous as we find our way in these appalling times, thank you.

1

In the beginnings: Introducing
Poetics of Deconstruction

After Jacques Derrida, 'I would like to choose words that are, to begin with, naked, quite simply, words from the heart.'[1] The direct flow of this sentence, metered by four commas, and issued with a cautionary conditional even as it shepherds words 'from the heart', closes the second paragraph of *The Animal That Therefore I Am*. It is apparently simple in that no 'jargon' is in immediate evidence. However, invoking the beginning several times in these opening lines, explicitly invoking the testamentary stature of 'In the beginning', Derrida circles around the desire for origins, the desire not to repeat, the desire for a 'time before time' by which means we should sense the sheer cascading scale of that which he calls the 'animal question' as it jostles with the concept and figure of nudity.

The very idea that words from the heart might convey anything other than unalloyed innocence might steer us at first, crestfallen, towards the deceit, lying in the wings, awaiting the reactive activation of the binary machine (either innocence or deception!). But this work of political philosophy cannot proceed without poetic means (and does not try to do so). Those poetic means might thus recall what appears to be the utter idiosyncrasy of another little text in which a hedgehog curls amongst the invocation that we should learn by heart. There, in response to the demand from the journal *Poesia* to answer the question 'What is poetry?' (a question posed to the first author published in each issue of that journal), Derrida addresses, 'you'.[2] That apparently direct address and demand for the presence of poetry – for it to be revealed in its naked truth, we might say, – do not result in a direct response or explanation. Rather, deconstruction has a taste for the oblique and this I have taken to heart. The oblique path responds to reasons both general and particular. Vis-à-vis the general, Derrida's work inherently cautions against the frontal engagement of 'Q&A' or classical forms of argumentation squaring off 'x' against 'y' since the very form of the question or of the argument will seek to predetermine what can be found. In this particular instance, the poetic itself is in question and it departs from both canonical formulations of genre and quotidian notions of poetry as the 'ineffable'.[3] Rolling two axioms up together, the poetic for Derrida is the invocation to 'learn by heart'.[4] This rolling bypasses the ostensible capture of synthesis to conjure an animal figure of vulnerability. Unlike its Romantic predecessor, this little hedgehog in the road is both curled up into a ball in upon itself *and* remains exposed to the world, unable to see death coming.[5] The two axioms take the desired organic and original spontaneity for naked words together with the demand for repetition and the committal to memory, to 'mnemotechnics' and 'a certain exteriority of the automaton'.[6] Rote repetition and the beat of the heart, the poetic – or 'poematic' – is both inside and outside, setting a rhythm that goes beyond any frame of opposition. Even that 'beyond' is freighted so that it does not signal the ascent of another form of transcendence but remains 'down there … humble, close to the earth, low down'.[7]

Summoning up some of the animals in his previous works – including the hedgehog – in *The Animal That Therefore I Am*, Derrida notes that 'almost all these animals are welcomed, in a more and more deliberate manner, on the threshold of sexual difference. More precisely, of sexual differences'. It is true that Derrida's now infamous nude scene before his cat generates a strong negative emotion, that of shame. We cannot, however, subtract that shame from the ontotheological conceptual history that would corral this cat to 'the animal' – identical to all the others, categorically lesser in kind than 'the human', while staining man's knowledge with the so-called original sin. What is singular about Derrida's work and why I draw so much from it is that he does not defend against his own, top-to-toe – or top-to-tail, implication in this history and in ways out of it. The threshold of which he speaks is one of welcome. The silkworms that Derrida observed as a child in Algiers do not issue an anxious lesson in sexual difference but a 'marvellous' secret secretion of sexual and animal differences. Not a word, not a snake. 'In the beginning', he writes – again – 'there was the worm which was and was not a sex, the child could see it clearly, a sex perhaps but then which one?'[8] This reverie from 'A

Silkworm of One's Own' risks a primal scene that does not congeal into the repetition of a single fault, exceptional signifier or compensatory fetish as per the normative account of psychoanalysis.[9] It thereby contrasts dramatically with the suite of ethical and political problems faced in *The Animal That Therefore I Am*, emblematized by the elementary armature of horizontal and vertical. This seemingly purely physical, or geophysical, axis accrues metonymic ambition when it automatically, art historically, stands in for landscape and portrait. Worse, the verticality of that portrait then becomes the metonymy of standing upright. While it is 'erection in general and not only phallic surrection' that is 'at the heart of what concerns' Derrida, as I explore these axes across a number of films, any resistance to divisibility on behalf of the concept or of the sign can be understood as a defence against the mutability of detumescence.[10] Deconstruction is improper. In an interview with feminist faculty at Brown University from 1984 – in the decade when women's studies programmes were gaining traction, Derrida muses that 'there is always something sexual at stake in the resistance to deconstruction'.[11]

That something takes perhaps its most unexpected form in Derrida's *Death Penalty* seminars. After his long lament for the lack of a philosophy of abolitionism amid the calculus of pain that is the anaesthesia of the death penalty, the first volume ends with resistance by means of the beating of Derrida's heart – and 'the grace of the other heart'.[12] That alone is arresting. But the next to last session of the second volume astonishes in the explicit and heartfelt alliance between deconstruction and feminism that it invokes. In that session Derrida returns to Sigmund Freud, following the red thread of blood as philosophy – and here psychoanalysis – shows itself to be unable to oppose the death penalty. Locating anxiety regarding the flow of blood, Freud finds himself turning his discussion of the defloration of women into one of female resentment born of penis envy (that is not the surprise). What is striking but, again, not surprising is Freud's twofold transition from the so-called 'primitive' peoples to his contemporary moment (itself a familiar synthesis from his colonial orchestration of the Subject of Europe in his speculative writings).[13] Firstly he locates the clearest instances of such resentment among 'the strivings and in the literary productions of "emancipated" women' in his own time.[14] Secondly, Freud risks a 'paleo-biological speculation' that roots this impoverished condition on their thwarted desire to urinate standing up.[15] With some restraint, Derrida responds that is not that Freud's 'targeting lacks insight' but that 'the phenomenon he has not failed to identify requires an interpretation about which psychoanalysis does not utter a word'.[16] With a heart-stopping divergence from the letter of Freud, Derrida aligns what he names the 'original and irreplaceable role of literature in the feminist cause' with the fact that it has been poets and writers generating abolitionist discourse – not philosophers 'or even politicians'.[17] In such a gesture he links it with his own writing, and the thought and the risk of writing in deconstruction in its broadest implication.

Resisting the direction in Freud that aligns moral rectitude and the rectitude or erection of the body standing before the law (or indeed a urinal), Derrida resists too the congenital figure of disability lodged in the logic of castration to which the resentful writerly woman is ostensibly destined. There is even a path emerging here that affirms the vulnerability of a resistant feminist emancipation with the 'nonpower at the heart of power' taking shape in *The Animal That Therefore I Am*.[18] That nonpower is at the beating heart of the transpecific living, and it is sexual without opposition.

*

In the beginnings that follow, you will find an accumulation of scenes given sustained attention. Circling around the desire for origins and for ends without being able to calculate either with the exactitude ordinarily taken for granted, *Poetics of Deconstruction* inhabits films, art and the psychoanalyses by which they might make sense other than under licence of the subject that calls himself man. That fiction of autonomy, therefore, subsides.

This book draws most substantially from Derrida, making the proximity of deconstruction to bodies of thought such as psychoanalysis more pedagogically available without that proximity being resolved into identity even as the conceptual groundwork that psychoanalysis shares with our inheritance of the dialectical tradition must be set aside. Yet the book brings to light commonality with a number of other writers. On the one hand, shared grounds emerge because Derrida was not in the business of branding concepts to be applied, willy-nilly, regardless of context. Rather he attended to the condition of the living in general even as it aims for singularity: this time, this space, this sex, this animal. On the other, this book engages those who have nourished thought in feminist philosophy, posthumanism and animal studies in ways that I find life affirming, most notably Hélène Cixous, Luce Irigaray, Cary Wolfe and Donna Haraway. In one of her lesser-known articles first published in 1987, written in response to the tall order of inserting the term 'gender' – or 'Geschlecht' – into a dictionary of Marxism, Haraway wrote, 'The evidence is building of a need for a theory of "difference" whose geometries, paradigms, and logics break out of binaries, dialectics, and nature/culture models of any kind.'[19] Yes, it is. Through your readerly progression to and fro across these chapters that unfold successively but also circle back, revise and resume, taking in primal scenes, death penalties, sacrifice, revenge, histories, autobiographies and apostrophes, perhaps that need can again take root.

2

The Animal Cure: Inhaling the other in *Dean Spanley*

In common with the gesture of reversal and displacement at the heart of deconstruction, Jacques Derrida asks, 'What, then, is *true mourning*? What can we make of it? Can we make it, as we say in French that we "make" our mourning? I repeat: can we? … are we capable of doing it, do we have the *power* to do it?'[1] This chapter begins around a dining table set to stage conversations between men: narratives of the adventures of male dogs double that scene. That may not immediately evoke the work of mourning, but soon, it will. The chapter explores the double prescription of what I call an 'animal cure' as it is suggested by the beguiling film adaptation of Lord Dunsany's 1936 novella, *Dean Spanley*.[2] *My Talks with Dean Spanley* does not self-consciously extend itself to support a politics or an ethics that would include animals; indeed, it remains close to the problems we readily associate with fables or allegory (in which animals habitually figure only as ciphers for human beings, as the 'beast' for human 'sovereignty').[3] However, by pushing this film in light of the work of Derrida especially as that work affirms a certain kind of psychoanalysis that cannot secure its principles as those proper to the human, I want to bring its more radical potential to light while acknowledging the problematic scenography that the film fields.

The animal cure in the sense that holds stronger narrative endorsement in this film is not for a sick animal, or animals in general, if there were such a thing.[4] Rather *Dean Spanley* enacts the 'talking cure' for melancholia as manifested in a cantankerous elderly man, Fisk (Peter O'Toole), by means of an animal. While the film self-consciously tells a tale of reincarnation – persuasively evoked through the cinematic convention of flashback, it is readily available to a conventional psychoanalytic understanding of the work of mourning as that which is best processed by enabling trauma to be put into words.⁵ In the canonical sense initiated by Sigmund Freud and Josef Breuer's early studies on hysteria, that distressing experience for which language has been unable to give voice coincident with its occurrence, but which lives on symptomatically, can only be abreacted by finding the 'words to say it' in a subsequent therapeutic environment.⁶ In this film, the unmourned deaths of his late son, Harrington, and his wife prescribe Fisk's extremely formal relationship with his surviving son, Henslowe. Meanwhile Henslowe (Jeremy Northam) becomes fascinated with the oddly convincing stories produced by the local clergyman – the eponymous Dean (Sam Neill), of his life as a dog when enjoying the scent of the rare Hungarian liquor, Tokay. Realizing that the dog, in whose name the Dean speaks, uncannily recalls the lost pet of his father's childhood, Henslowe brings his animal cure into effect not by means of an actual psychoanalytic session but through reminiscences nonetheless, here provoked at the scene of a dinner party. From the moment that this pet, Wag, is 'returned' through the medium of the Dean's apparent recollections, Fisk can begin to cry and thus to admit grief. Yet from this moment too, the intoxication with Dean Spanley fades: the normatively satisfying resolution of the last scene suggests a newly happy Fisk secured by a new pet dog.

Dean Spanley makes a series of doubles between humans and dogs: son and dog (Harrington and Wag), dog and Father (in the Dean and also in Fisk) and also of dog friends and human friends (Wag's doggy friend and Wrather the 'conveyancer' (Bryan Brown), Henslowe's fellow conspirator in the supply of Tokay).⁷ It self-consciously does this with the key scenes of the film too – men assembled around a dining table/dogs running through fields. By convening the entwined narratives through a ritual meal, metonymized by Tokay, *Dean Spanley* invites reflection on the primal feast and the legend of consanguinity between human clan and totem animal as invoked by, or indeed claimed by, Freud in 'Totem and Taboo'.⁸ The cannibalism of the primal feast is one that initiates the possibility of representation along with its initiation of law and sociality, according to psychoanalytic and anthropological legend. It is thus an 'origin story' that is also bound to a firm end. As is perhaps suggested by the insistence on fathers, sons and brothers in the film chosen to open this book, this foundational representation is precisely of *patri-arkhē*: the Father at, and as the origin that is also the necessary destiny of, the law. *Dean Spanley*, in foregrounding the domestic ingredient of the dog as man's best friend, enables discussion regarding mourning among humans and animals. However, rather than maintain the film's loyal proximity to totemism, in Freud's restricted sense, this chapter brings Derrida's work to bear upon it.[9] This involves not simply rethinking what it is to mourn and whether animals can be mourned as such but treating psychoanalysis as that which must itself be cured of the regime of representation that it so frequently endorses. Deconstruction does not frontally oppose psychoanalysis, since any such oppositional gesture remains locked into that which is opposed. Rather, by means of the ethics of what Derrida names 'eating well', it traces another 'cure', one that will shake up the hard-and-fast distinction between 'the human' and 'the animal'.

Totem and Tokay

Most of the still proliferating commentaries on *The Animal That Therefore I Am* concentrate on Derrida's encounter with the animal in or as his deconstruction of the persistent philosophical support for human exceptionalism.[10] Yet observant readers will have noticed that, in reference to his own 'zootobiography', Derrida remarks that his writings have 'welcomed' animals on the 'threshold' of sexual difference.[11] 'More precisely', as he immediately follows, 'sexual differences'. I take this gesture to index the fall of the concept and the affirmation of differences unmanaged by dialectics (a form otherwise inherited and maintained by psychoanalysis). The word 'welcome' draws attention to an ethics of hospitality to the other, rather than a manifesto of rights: Derrida's transfigured autobiographical texts welcome sexual and animal others without their advance being disclosed in advance.[12] While this kind of welcome includes the complication of being hostage and not simply host to unknown others (he remains indebted to Emmanuel Levinas in this regard), Derrida nevertheless offers a scene of hospitality that moves away from canonical autobiographical and philosophical negation or abjection of those others in the name of subject that calls itself man.[13]

The scenes of hospitality that structure *Dean Spanley*, however, echo these problematic processes of negation or abjection, not least in regard to the primal feast that Freud deduces must have occurred at the origin of culture.[14]

Drawing on numerous anthropological sources, Freud positions this feast as a ritualized exceptional event that permits the clans of so-called 'primitive' cultures to kill and to eat their otherwise protected totem, a specific animal with whom they assume a consanguineous relation (sexual reproduction being unrecognized). Without this ritual, such a meal would have been wholly taboo. As a codified and momentous event, the ritual both breaks the law against parricide and founds it. Drafting 'Totem and Taboo' into the service of the Oedipus complex – the only factor that can offer persuasive explanatory force to the sheer dread of incest evidenced but not explained by anthropological acknowledgement of the universality of the incest taboo – Freud draws in the present of his clinical observation of animal phobias. His modern phobics exhibit ambivalence – that is both love and hate – towards their feared animal, and thus Freud finds continuity between primitive and modern cultures in support of the theory of psychoanalysis: 'In every case it was the same: the fear was basically of the father, *where the children under examination were boys*, and had merely been displaced on to the animal.'[15] Regardless of any doubt raised by the absent question of girls, the primal meal requires greater finesse, and Freud further entrenches the father at the origin of culture by supplying a revised wish for which the primal feast is already a dilution. The consanguineous imagination of shared blood is of no consequence: our animal ancestry is merely a displacement of patriarchy. Freely evoking Charles Darwin (without any actual reference), Freud imagines the overcoming of this primal father by the 'band of brothers' who murder and eat him.[16] Such is the enormity of their guilt that the father is resurrected in name and in/as law, without even having to die since the wish to despatch him would have been a force enough for psychic reality. As feminist scholars such as Kelly Oliver and Elissa Marder have remarked, 'Totem and Taboo' glosses over both modes of kinship that predate the nuclear family as well as the scattered incoherent references to feminine fancies and maternal deities in the rush to render the father original, necessary and human.[17] As scholars such as Deborah Bird Rose demonstrate, totemism cannot be confined to the European imaginary of a primitive past but lives on in divergent ways in indigenous modernities.[18]

Retaining the notion that affective response to criminal events found culture as law, Julia Kristeva not only invokes the murder and cannibalism of the father but also reminds the readers of *Powers of Horror* that incest with the mother drops away from Freud's attention by the end of 'Totem and Taboo'.[19] Most of the literature following Derrida on the question of the animal has remained within the philosophical terrain that he indicates, targeting the Cartesian legacy of persons such as Heidegger, Levinas and Lacan, yet Kelly Oliver has shown that female thinkers such as Kristeva also demand to be rethought in light of the human exceptionalism that they too legislate. Thus I introduce her with caution, peripherally in the case of *Dean Spanley*, but more frontally and in connection with texts in addition to *Powers of Horror* in the following chapter. In Kristeva's case, as Oliver points out, alongside the human and masculine route to language, the abject haunting of any border does not only recall the body of the mother – through 'our personal archaeology' – but also, on a wider scale, it recalls the animality that was otherwise expelled as 'representative[s] of sex and murder', or lawlessness.[20] For Kristeva, we might say that where animal and sexual differences traverse the same horizon, they do so as a threat that cannot be accommodated to human sociality – and yet that threat is itself moderated by the return of the singular concept of difference.

Kristeva might address Freud's notorious blind spot regarding femininity, but she does not offer a feminist counter model (as she herself acknowledges). Moreover, Kristeva endorses the requirement that the social rest upon the exchange of women between men, indexing the symbolic exchange of signs, for fear of the untutored lawlessness of the mother and the hunger of her animal bite (this problem in her work is explicitly addressed in the second chapter of this book).[21] While the figure of the mother is scarcely in evidence amongst the homosocial cast of *Dean Spanley*, the liminal nature of abjection means that her direct representation is not the issue.[22] Given the encoding of the scene of the meal as both paternal and fraternal in Freud and this film, Kristeva provides a useful supplement through her attention to the abject power of particular substances. Signally, in *Powers of Horror*, food as that which traverses the mouth threatens the sociality of man; food as an 'oral object' recalls the archaic relations between human and m/other.[23] 'The body', she says, 'must bear no trace of its debt to nature: it must be clean and proper in order to be fully symbolic'.[24] Thus food can always 'defile'.[25]

Having set the table of this chapter with the spectres of cannibalism and incest, I want to turn to *Dean Spanley*. Thursdays table a dry ritual between Fisk and Young Fisk (as Henslowe is schematically addressed by his father).

'Young Fisk' arrives at his father's house and they address matters of fact, untouched by affect. Henslowe himself ironically refers to their scheduled meetings as rituals, and ones that he wishes were 'dismantled'. An altogether more fascinating ritual transpires for Henslowe with the Dean. Underlining the displacement of father for Father, Henslowe arranges his meetings with the Dean on Thursdays. Aware of this substitution, Fisk makes his own: when they do manage to get together for a (Thurs)day out, Fisk pointedly trips up a young boy (i.e. in lieu of Henslowe).

At first, procuring the Dean's favourite liquor is simply to facilitate their meeting and to allow for the Dean to expound upon the unlikely topic of reincarnation (one Fisk characteristically dismisses as 'poppycock'). Almost immediately the Dean is implicated in that very topic as his unusual degree of pleasure inhaling Tokay – the script positions him as 'entirely focused in his nose' – leads him to wish for the 'olfactory powers of the canine'.[26] More disconcertingly, as he continues with increasingly outré remarks, his first-person narration becomes uncannily canine. He does not mimetically sound like a dog; rather, his sudden marked interest in cats, smells and the love of a master evokes the point of view of a pet dog. At this early stage in Henslowe's intoxication with the Dean, no images flesh out his story as a flashback in the manner cinema habitually treats evidentially as memory. We have to take his word for it and Henslowe's fascination for our own. The clue to the change in perspective within the Dean – from human to canine – comes through an unusual comparison. He opines that

> to pull a dog away from a lamppost is akin to seizing a scholar in the British Museum by the scruff of his neck and dragging him away from his studies.

Making kin of the inhalation of urine and the study of books threatens the clean and proper body (inhalation of urine is not named as such, but the comparison follows swiftly on from the Dean's appreciation of the Tokay consolidating their metonymic connection and collapsing the latter's ostensible remove from more abject fluids).[27] In this contagious comparison, dog and (human) scholar are made of the same stuff, and up to the same activity. A dog reads traces of urine like a scholar reads writing.[28] By implication, to urinate is to write (to leave a trace, and one vulnerable to erasure); to smell is to read. Metaphor assumes that the meaning of the term of comparison anchors that to which it is compared. Here, however, the scholar is already dog-like, seized by the 'scruff of his neck'. Later, in the film's climactic sequence, it is Fisk that makes a similar comparison in which his wife calls him away from reading Balzac: 'rather like dragging a dog away from a lamppost'. In both cases there is no mention of the word 'urination', which is tidily metonymized by the lamppost. Even as metonymy – a relation of comparison based on proximity – the abject contact between urine and study is finessed. In the latter scene the Dean describes the pleasures of eating a whole rabbit, fur, bones, guts and all, waxing lyrical about the smell of fear. Although by then we are regularly treated to visual flashbacks of Dean Spanley in the guise of Wag the spaniel, learning about the world from his roguish mongrel friend (clearly meant to be Wrather), this visceral desire for the fear-drenched rabbit is overheard by Mrs Brimley, the housekeeper (Judy Parfitt). Literally peripheral to the proceedings, her mortification is presented as comic. She hears something that she should not and cannot understand, unaware that she is listening to the Dean as a dog. Dean Spanley is at pains to make sure that *our* guts are never turned (as we, audience, metonymically join with the enraptured homosocial circle of Henslowe, Fisk and Wrather). While Mrs Brimley has prepared the food (and insisted on preparing something more special than the 'hotpot', to which Fisk habitually constrains her), this is not the meal at stake for the assembled men. Indeed the film gives scant attention to any food as such, in spite of such protracted attention to the *mise-en-scene* of dining. That they eschew the tradition of leaving the table in order to enjoy port in separation from any 'ladies' that might ordinarily be present to remain at the table confirms which meal is in focus. They partake of the *story* of downing an entire rabbit mediated by the aroma of Tokay in order to share in the memories voiced by the Dean.

Unable to be seen, smell is elusive. It lends itself to the uncanny tale of *Dean Spanley*, posing the unfathomable question of whether the Father was once a dog, while the domestic status of that dog points back to Fisk (again containing the impure legend of consanguinity).[29] The film supplements smell's invisibility with the Dean's rhetorically exaggerated appreciation of the Tokay. This rhetorical exaggeration is given clearest visual expression in the final dinner sequence. There, in close-up, the Dean raises his glass to his nose, reminiscing about the delicious smell of fear, the classical soundtrack swells and the film cuts to the comedically rapid appearance of sheep being chased

Figure 2.1 Inhale (*My Talks with Dean Spanley*, 2008, Dir. Toa Fraser, UK/New Zealand).

over a hill by dogs delirious with olfaction. Becoming virtually airborne in their haste, the white clouds of leaping sheep conjure their own scent. In his discussion of smell and Freud, Akira Lippit refers to its paucity of visible trace as an immateriality that bars smell from forming a 'semiotic system'.[30] In this view a scent could never form a sentence. Need we, however, be so quick to assume either that smell is immaterial or that materiality secures signification.

Tokay is elusive. Wrather, the 'conveyancer', sniffs it out, squirreled away in the wine cellars of the wealthy, though he soon dispenses with a finder's fee for the sake of a place at the table. Tokay is not disgusting. Even if it is rather syrupy, it is not presented as abject. One does not even have to bother the mouth by drinking it. For the purposes of *Dean Spanley*, Tokay is taken by nose. Intoxication with Tokay is not coarse inebriation. This rarefied liquor is celebrated as ostentatiously cultural, even with aristocratic connotations given its origins in the then Austro-Hungarian Empire. Rather than confirm human desire over animal need, the Dean imagines that a dog might appreciate its aroma all the more. Perhaps the ritualised, exceptional consumption, the elevated palate required to appreciate Tokay protests too much and defends against the possibility that pollution inheres in food. For Henslowe and Wrather this liquor is instrumentally the vehicle for the Dean's transport. Fisk blunts the allure of the Tokay not by emphasizing disgust but by dismissing it as nothing more than its source components: 'fermented grapes'. Outright disgust would too easily register the psychoanalytic mode of repression. Freud famously narrates – albeit in a footnote itself banished to the bottom of the pages even as its girth takes up most of those pages – the vertical elevation of man as coterminous with the predominance of the sense of sight, with both verticality and visuality set against the horizontal and olfactory order of the animal.[31] Closer to the earth, closer to the sexual and excretory organs of other four-legged animals, this plane is one foregrounding the sense of smell.[32] Freud even singles out the dog as both a 'faithful friend' and one whose name is appropriated as a 'term of abuse', since, he says, 'it is an animal whose dominant sense is that of smell and one which has no horror of excrement, and that is not ashamed of its sexual functions'.[33] Defending against a disgusting smell then bespeaks the desire for the sexuality and the animality it indexes. The Dean's elevation of Tokay might be read in this context, especially given the homosociality the dinners also convene, eliminating women and cultivating men – and male dogs. Yet for Fisk, Tokay occupies no extreme: it is neither disgusting nor wondrous. In common with his reduction of Mrs Brimley's culinary repertoire to the economically descriptive 'hot pot' and his curt reduction of things that have 'gone to the trouble of happening', including the deaths of his wife and son, as 'inevitable', Fisk dampens social engagement until he recognizes his dog in the Dean.

Scents and sentences

In the material already introduced we have a mounting sense of the sociality at stake in the consumption of food in excess of a supposedly simple nutritional need. Freud has laid out the primal feast as a scene in which animality is exchanged for (human) paternity, a greedy paternity that also founds the law, culture, history, etc.; Kristeva indicates the feminine as well as animal territory mapped by the mouth that also haunts this feast. It is Derrida that names an ethical imperative that opens upon all the senses in general: 'one must eat well.'[34] We must be clear this 'eating well' does not equate to fine dining or good manners. Rather the 'good' (underlined by his translator's emphasis on the original '*bien manger*') speaks to an ethics that for Derrida cannot be resolved into a calculable formula. Sara Guyer notes that '*un homme de bien*' is a not merely a 'good' man but a man of property and that '*bien*' is connected to the Greek '*oikos*', drawing together 'the home, ... the "proper", ... the private, ... the love and affection of one's kin'.[35] Not only are we always in a relation of 'eating the other' and being eaten by them but, Derrida tells us, that the ingestion the verb indicates is limited neither to food nor to intake by mouth: 'For *everything* that happens at the edge of the orifices (of orality, but also of the ear, the eye-and all the "senses" in general) the metonymy of "eating well" [*bien manger*] would always be the rule.'[36] In the 'Eating Well' interview conducted by Jean-Luc Nancy, Derrida himself asks, 'What is eating?', having so expanded this ostensibly self-evident basic need, now re-posed as the 'metonymy of introjection' and as a question carrying philosophical weight.[37]

Contiguous with eating, introjection names the psychic process of identification and itself metonymizes the work of the psychoanalysts Nicolas Abraham and Maria Torok, from whom Derrida implicitly draws, albeit in a modified fashion.[38] For Freud, Abraham, Torok and Derrida, we must 'eat the other' if we are to form our own ego; that is to say, our earliest identifications with others occur as a form of ingestion that we are obliged to swallow. Yet Derrida departs from psychoanalytic canon: the 'must' here refers to an ethics of infinite hospitality – one takes in the other but does not decide which other. At the same time there is a 'cannot' in that we cannot measure or decide how much of that other to take in: the critical interface of literal and figural ensures that we cannot totally appropriate the other through this ingestion. That is not the departure. But that the ostensibly physical practice of eating and ostensibly psychical process of introjection may be said to share a border does not only point to the difficulty of forming a clear succession or separation between literal and figural; it consequently also points to the difficulty of distinguishing between need and desire – themselves synonyms for nature and culture – and thus, for Derrida, if not for Abraham and Torok, between humans and other animals.

Departing from the metaphysical conceptual path that orders and interlinks these terms leads Derrida to pose the ethics of the 'One must eat well' as offering an 'infinite hospitality'.[39] This infinite hospitality strikes at the 'carno-phallogocentric' heart of metaphysics in calling into question the structure of sacrifice that it conserves.[40] This mouthful of a term brings Derrida's existing critique of the conceit unifying the presence of the word (logocentrism) with that of the phallus (phallogocentrism) into contact with a sacrificial logic of transubstantiation.[41] It arises in the interview with Nancy by virtue of the title of the journal issue, in light of which frame 'Eating Well' takes place: 'Who Comes After the Subject?' While the topic of 'the Subject' is nominally in question, Derrida finds it reinstalled in the maintenance of the 'who' and still levelled against a 'what' of lethally lesser status. In contrast, Derrida refuses 'to see the "who" restricted to the grammar of what we call Western language, nor even limited by what we believe to be the very humanity of language'.[42] Even ethical thinkers with whom Derrida shares ground such as Emmanuel Levinas fall foul of the configuration of sacrifice. While a 'Thou shalt not kill' may be invoked, even as a first principle, Derrida draws attention to the way in which killing is managed such that a 'non-criminal putting to death' symbolically and legally distinct from murder is still set aside for some beings.[43] (I return to the linked problem of the death penalty as such in the last two chapters of this book.) This Levinasian ethical law implicitly addresses a human community, for whom the killing of non-humans does not count.[44] Explicitly affecting those we call animal, the sacrificial loophole for legal killing can and has been turned on humans *called* animals, often through pestilential figures such as 'vermin' (notoriously in anti-Semitism and especially in the context of the Holocaust, but that ugly self-justifying rhetoric of pestilence that has returned to current political events at the time of writing with particular regard to immigration). As Freud describes (in 'Totem and Taboo'), so Derrida critiques this community, which, moreover, privileges brotherhood: the virility associated with the Judaeo-Christian carno-phallogocentric subject is indeed that of the 'adult male, the father, husband, or brother' demanding a sacrifice.[45]

The contiguity between eating and introjection provokes another conceptually challenging question: 'In what respect', Derrida asks, 'does the formulation of these questions in language give us still more food for thought? In what respect is the question … still carnivorous?'[46] The carnivory of the question is given with the caveat 'formulation' 'in language'. While Derrida himself does not flesh out the full archive of thought being redrawn here, his question recalls the Freudian understanding of language acquisition as the substitution of breast for word: in the crossover between the metaphysics of presence and psychoanalysis, a suite of metonymies, milk, breast and mother, all bound to the psychoanalytic fantasy of satisfaction, give way to the substitution of language.⁴⁷ However, that retention of meat for word is given an uncanny breadth if his invocation of all the senses in general is brought into contact with another analytic frame, that of Melanie Klein. While Klein is only fleetingly referenced across Derrida's corpus, we cannot miss the reciprocity with her 1936 text on 'Weaning':

> the child receives his main satisfaction through his mouth, which therefore becomes the main channel through which the child takes in not only his food, but also in his phantasy, the whole world outside him. Not only the mouth, but to a certain degree *the whole body with all its senses and functions, performs this 'taking in' process* … the child breathes in, takes in through his eyes, his ears, through touch and so on.⁴⁸

Perhaps it is here that 'Klein perhaps opens the way', the highly elliptical point on which Derrida finished his first essay on Freud.⁴⁹ Where Levinas poses the face as that which says 'Thou shalt not kill', Derrida displaces the humanism that the face proposes not just with the mouth but with all the orifices, thus weakening the association with speaking subjects.⁵⁰

Rather than legislate anew, invoking a new, improved, law on which we could always rely, the Derridean ethics of infinite hospitality keeps the question of what it is to eat well open. Refusing to sequester symbolic anthropophagy as a human practice distinct from literal cannibalism committed by the untutored, animals, those who lack the law, Derrida implies that vegetarians also 'eat meat' in the place where eating and introjection touch.⁵¹ Harking back to my remarks on early identification as a form of 'eating the other', there is a metaphoric carnivory at stake that is not definitively refused by the practice of a vegetarian diet. As we are starting to gather, the place where eating and introjection touch may vary! Thus it is not necessarily organized linguistically, and its implications may stretch beyond humans, including those who practice a vegetarian or vegan diet. To take seriously this demotion of the mouth as the ostensibly privileged path to introjection, a process that finds its due outcome in speech, is to invite the question of non-human ingestion, non-human carnivory and thus non-human modes of signification.

In their work on mourning, framed in binary combat as 'Mourning *or* Melancholia: Introjection *Versus* Incorporation', Abraham and Torok distinguish these processes in ways that lend themselves to thinking about

Fisk's abrupt dismissal of pain.⁵² In Derrida's 'Foreword', called '*Fors*', for their book *The Wolf Man's Magic Word*, he warns against the 'limitations' of a 'linguisticistic' reading of their work, one easy to make since it stems from the very 'base of the[ir] enterprise'.⁵³ This reading overdetermines the mouth as the privileged oral locus of '*verbal* language', one whose presence fills the gap left by the breast.⁵⁴ Speech comes first, and speech is presence (the metaphysical problem inherited by psychoanalysis). Derrida underlines the inadvertent fracture in this logic: the substitution is '*partial*'; presence is a '*figure* of presence'.⁵⁵ Psychic life is in mourning from the start.

Abraham and Torok differentiate mourning and melancholia through two different relations to the literal and the metaphoric. Rather than introject the lost other as a metaphor, the melancholic incorporates that lost other as an object that thus refuses metaphoricity.⁵⁶ Melancholic incorporation involves the fantasy that one eats this object precisely '*not* to introject it', as Derrida puts it, 'in order to vomit it, in a way, into the inside into the pocket of a cyst'.⁵⁷ This 'cyst' is the secret 'crypt' in Abraham and Torok's terms (one that we should read in a *linguisticistic* manner): the one for whom the melancholic fails to mourn is squirreled away within it. Abraham and Torok oppose the withheld path of incorporation to the sociality of introjection. For them, 'Introjecting a desire, a pain, a situation, means channelling them through language into a communion of empty mouths' (empty by virtue of the process of weaning).⁵⁸ As Derrida writes, in summary of Abraham and Torok, 'Introjection speaks … Incorporation keeps still, speaks only to silence or to ward off intruders from its secret place.'⁵⁹ This crypt of language depends, for Derrida, on the logic of a primary substitution for the maternal breast configured as presence. Language, cryptic or otherwise, is here caught in the logic of re-presentation. Of course Derrida gives emphasis to the supplemental nature of the substitution of breast by word: supplemented, the breast loses the sense of an originary completion (without thereby falling into a logic of lack). Rather than the full presence of the breast, metonym of the mother's body, metonym of nature, Derrida posits an original writing: general 'hieroglyphia' precedes possibility for thinking the crypt.⁶⁰ This does not push the supposed ground of 'nature' further 'back' to a more secure place (in which it remains inert and precisely ahistorical) but rewrites it as writing already.⁶¹ Thus the general writing of nature also disperses the singular path to language as the human response to lack.

Ingestion that does not necessarily pass by way of the mouth immediately evokes the nose for Dean Spanley, as well as the ears for his audience, while the crisis in language summons Fisk.

Pet seminary

Fisk is blunt. He neither 'wastes' words by indulging their figural capacities nor worries about offending others. The congregated guests around the dinner table in the climactic sequence are at first beholden to his stories, ones they have not come to hear. We hear how his late wife dragged him from Balzac to aid their two sons, out on a rowboat on a stormy Lake Windermere.[62] Mocking her fears, the cantankerous Fisk addressed the storm intoning 'Give Up Your Dead!' as if they were already deceased. Fisk's disregard for emotional responses evidently predates the death of Harrington (fighting in the Boer War, his body never recovered). At dinner, once the Dean has again become the focus of attention, we learn the incorporative extent to which Wag and Harrington share the same fate, both marked by a 'non-criminal putting to death'.

It is the Dean's desire to remain at the table that again prompts the olfactory metaphor spurring his uncanny reflections. Leaving the table would be equivalent to having a bath 'when one ha[d] just gotten comfortable in one's smell'.[63] Bodily, animal, smell is thus brought into proximity with the bouquet of Tokay as a form of clothing, troubling its primary horizontality in Freudian legend. Bathing, cleanliness, leads to the embarrassment of nudity (I return to the question of nudity in detail in Chapter 5). The *séance*-like scene in the dark environment of the book-lined room housing the dinner resumes. Or, in Derrida's neologism, the '*animalséance*' resumes: Leonard Lawlor unpacks this term as both 'animated impropriety' and as a 'session of the animal' (session having both a psychoanalytic and an occult implication).[64] Fisk is astonished. Before he can issue an insult, the Dean resumes his otherworldly discourse. Speaking from the twinned crypt of Harrington and Wag, he makes casual reference to being called Wag by the Master. Fisk is transfixed. The Dean's ensuing stories entrance Fisk even more than Henslowe, and in transferential style, he soon responds as the Master in question, even recognizing himself as one who administered an occasional beating to Wag (to the raised eyebrows of Henslowe).

The tales to which Fisk is party bring the whole group together. Here we gain a clearer picture of urination as a writing practice, of the enticing smell of fear and of friendship between dogs (the 'unmastered', unnamed stranger and Wag, domesticated, his species loyalty divided by a love of the Master). This picture is fleshed out by luscious flashbacks cinematically coded as first-person memory in that they are attached through successive sequencing to the Dean but shot from a low angle, from a dog's eye view. The latter gives credence to the Dean's story and draws those who see these sequences – the cinematic audience – into the film through that canine viewpoint, making dogs of us all: exuberant dogs often taking up the whole frame, dogs in the prime of life, sometimes with a slightly self-consciously comedic feel produced through a slow-motion close-up of wind in their coats, all suggesting that yes, those times were fantastic.

Fisk is particularly taken with the Dean's assurance that to find home, after running unfettered through farmland with his pal, he had only need to turn towards it.[65] This confidence mystifies Fisk since Wag had disappeared, like Harrington, and no body had been recovered. Yet the dogs do not arrive home, since, as the film shows while the Dean cannot tell, a farmer shoots them dead. Fisk is rapt. As he stares at the Dean, the scene cuts back to that same field, shot in the same light, but this time with his son Harrington riding a horse

Figure 2.2 Exhale (*My Talks with Dean Spanley*, 2008, Dir. Toa Fraser, UK/New Zealand).

across it. With the sound of gunshot, the scene cuts, and we see Harrington lying dead in the field as the Dean narrates Wag's last thoughts of 'home in [his] heart and the master waiting. No, no pain'. The Dean's audience are visibly affected (indeed it would be hard to remain unmoved). Fisk, weeping gently, touches the Dean's hand affectionately. With new consideration for the feelings of others, Fisk retires saying that he is 'put in memory of Harrington', the son whose name he uses for the first time in this film. Finding him crying in the hallway, the astonished Mrs Brimley asks Fisk if he is alright. 'He was shot', he replies. But this time the direct statement reveals his pain and opens the crypt.

The personal pronoun is ambivalent as to which death it refers, that of Harrington or Wag. Both shot: the dog as an animal trespassing on a farmer's land and as an animal that can be killed without criminal offence, indeed without truly 'dying', merely perishing according to Martin Heidegger; the son as a soldier, engaged in the lawful practice of killing those designated 'enemy' is himself so killed, a casualty of war.[66] The Dean's apparent recollection gives a representation to the traumatic absence of any such for Fisk, and one that affirms 'no pain'. In contrast to the formerly inexplicable disappearances of Harrington and Wag, Mrs Fisk died of grief for her son, in emotional pain 'enough for both of us', in Fisk's encrypted opinion. Yet the film shows no engagement with Fisk's grief for his wife – who remains nameless, only his belated double mourning for son and dog.

Figure 2.3 Cry (*My Talks with Dean Spanley*, 2008, Dir. Toa Fraser, UK/New Zealand).

'Eating' Wag as metaphor (by taking in the Dean's narration) allows the name of Harrington and sociality to resurface.[67] Talking now with uncharacteristic familiarity, Fisk hugs Henslowe, calls him too by name and volunteers to see him next on any day of the week. 'One moment you are running along, the next you are no more', a tearful Fisk utters, with the pronoun again lending ambivalence to its reference. Substitutable, the second person could indicate Henslowe, Harrington, Wag, Fisk himself or any other.

With the animal cure pronounced and Fisk returned to sociality and/ as paternity, fascination with Dean Spanley fades: this Father too has been figuratively consumed. Henslowe next finds his father not ensconced in the parlour but outside, playing with a young spaniel.[68] A dog has replaced the Dean. A dog comes home and 'home' is returned to its orderly self. Watching Henslowe watching his father, the film frames Mrs Brimley next to the painted portrait of Mrs Fisk. Mrs Fisk, nominally *the* maternal figure in the film, is never mentioned in Fisk's restored sense of feeling but is nevertheless symbolically assembled through this representation with the group approving Fisk's joy in his new pet.[69] In the spirit of doubles dogging this film, Mrs Brimley metonymizes the maternal – but a maternal already in service to the father/law. Employed as the housekeeper, she is bound to maintain clean borders rather than threaten their collapse in Kristevan abjection (the picture frame enclosing Mrs Fisk might also be read in this light). Later in the film, talking to her late husband in the form of the chair in which he used to sit, Mrs Brimley refutes the idea that she would ever cook anything so disgusting as a whole rabbit.[70]

Is the new spaniel a substitute for Wag or Harrington? Maintaining totemic ambivalence of whether humans and animals are distinct or consanguineous, Henslowe's closing voice-over suggests that reincarnation might be something to greet with anticipation, and, that should he be reborn as a dog, he hopes to belong to a 'master as kind as [his] father'. Given that Fisk had affirmed that he beat Wag (only) when it was necessary, and the Dean had spouted the colonialist view requiring the colonized to love their colonizing masters – characteristically confusing servant with dog – this wish too remains thoroughly ambivalent.

What is clear, however, is that not any animal could induce this cure for Fisk. I have indicated that the animal in the Dean is domesticated rather than wild, indexing Fisk rather than unleashed animal others. The film also deliberately repudiates felinity. The Dean reviles cats, berating their lack of understanding of the sport of the chase, and Swami Prash specifically expels them from proximity to man (the generic 'man' is categorically specific) early in the film. Speaking of reincarnation at the event that first brings the protagonists together, the Swami vehemently rejects enquiries after the possibilities of a feline soul made by women in the audience. In spite of its scenes of hospitality, *Dean Spanley* does not welcome animality; rather, its feminine taint and concomitant disrespect for (the law of) the master is held at bay while the film maintains a domesticated totemism commanding masculine descent. Derrida asks what would happen to fraternity should an animal – or a sister – enter the political sphere.[71] *Dean Spanley* splits between negative and affirmative readings: the symptomatic containment of the animal precisely as man's best friend, absorbing the dog within the discourse of friendship and ingenious pointers to deconstructing the conceptual hierarchy of man and animal.

Laurence Rickels ascribes the role of inoculation against death to the pet.[72] A loyal Freudian, he means specifically paternal death (the primal feast is lived every day). Prescribing carno-phallogocentrism anew, Rickels posits the eating of meat as that which develops resistance to the pain of loss.[73] Eating meat is indeed an 'animal cure' (but this time as food preservation). If the pet's death is unmournable for Rickels, this is because this classical traffic in substitution is one way (pets rehearse human death but nothing and no one does so for them). Rhetorically maximizing his own ambivalence regarding pet death, Rickels refers to 'cut[ting] their losses with the paternal economies of sacrifice, substitution, and successful mourning'.[74] Whether this means breaking from or mixing in with such economies, the prospect of successful mourning brings us back to Derrida and to *Dean Spanley*.

The trace of a cure

I began this chapter with Derrida's question regarding whether 'we' are 'capable' of true mourning. This phrasing resonates with his deconstruction of the habitual framing of human response versus animal reaction.[75] In *The Animal That Therefore I Am*, rather than simply extend the ability to respond to animals, Derrida questions the way in which ability is construed as the proper of the human (the ability to speak, respond, reason, etc.) and proposes a 'weak ability' in the common question 'can they suffer?' I return to the difference upon which Derrida insists by means of this alteration in emphasis in Chapter 5. In this chapter the question remains do *we* have the *ability* to mourn?[76] Were 'ability' to be classically adhered to, then any sense of the unconscious would be betrayed (affected, as it surely must be, by desires that my conscious mind would disown). Such is the pull of the conceptual field from which 'ability' derives that even psychoanalysis struggles to disengage. Derrida troubles the binary confidently asserted as 'Mourning *or* Melancholia', by Abraham and Torok, a division that circulates the one for whom we 'successfully' mourn and encrypts the one for whom we fail to do so.

The last chapter of this book becomes obliged, by the focal demands of *White God*, to attend to autopsic architectures made theorizable in light of Derrida's seminars on both *The Death Penalty* and *The Beast and the Sovereign*. The violence of the law overseen by the latter figures is obscured by the ostensibly genteel environment of the fable – the fabulous work of Dean and dog in *Dean Spanley*. But a rare sense of autopsy secures them. In addition to its regular usage as the medical practice of the dissection of a corpse, autopsy names 'intimate commerce' with the gods: it thus speaks to the totemic, to participation by means of ingestion.[77] The eleventh session of *The Beast and the Sovereign* provides a scene a good deal more dramatic than that of Spanley by virtue of exorbitant scale and insistence on the primacy of absolute vision: a King oversees the 'inspectacular' autopsy of an elephant.[78] This intimate commerce is overwhelmingly signalled as a visual ingestion, for which the tools of the autopsy are merely the herald. The séance of *Spanley* may open the route to an olfactory ingestion, but that opening is obscured in so far as the two scenes, the dinner table and the dog's adventures, conscious and unconscious, are resolved by the narrative production of the (equivalent) missing bodies of dog and son and secured in the cure of Fisk from his melancholia. Yet in the course of this traffic in bodies and the imaginary light of the Sun King, Louis le Grand, Louis XIV, illuminating and thereby governing his court and the body of the world's largest land mammal, a traffic that is seemingly one way and reflective only of the enduring autopsic vision of this absolute sovereign, Derrida opens up the *cura* of curiosity:

> The culture of curiosity thus organizes the showing of curiosities for curious crowds, but the same culture of curiosity also had ambitions to treat, to care for, if not to cure. Or even to liberate by locking up differently. The *cura* of

this curious curiosity always hesitated between two forms or two aims of what is always a *treatment*.[79]

The animal cure of *Dean Spanley* patently cares about shepherding a more loving relation between fathers and sons (and, on the face of it, who would tarnish that aim?). Carried by scent as the trigger of memory, the film has invited us to think of dogs as beings who write and humans as those who may be clothed by smell. In so doing it approaches curing philosophy and psychoanalysis of their persistence in dividing the human from the animal. But that new proximity, that intimate commerce of ingestion, is defended against by means of the masculinist ruse that fetters its departures from the discourse on 'the animal': when dogs are pulled back from writing – with urine – this is at the hands of a female figure whose action is tantamount to toilet-training; elements that might usually impart abject revulsion – sniffing urine, eating entire rabbits – and thus bespeak the defilement of the Kristevan mother, are 'cleaned up' and elevated to ritual events. *Spanley* even ends with the son's expressed desire for a good Father who will treat him, and even (lawfully) discipline him, like a pet dog, endorsing classically satisfying narrative closure. Rather than remaining attached to a virility in which man's best friend is indeed like man, his fellow man, we must risk the insecurity of following through on Derrida's insight that the general condition of writing affects the 'living in general', cultivating, as further chapters in this book will show, 'differences that grow'.[80]

3

Raising animals: Between the basement and the kennel in *The Woman*

She bites off his finger. She bites off his ring finger. She swallows the flesh but spits out the ring. Her first gesture when realizing her new situation as the restrained captive of the man who abducted her is to violently refuse the very sign of a hetero-patriarchal economy of exchange. It is not worth eating.

Teeth should not be this sharp; mouths should not be this filthy – it is not right, surely. The dreamlike title sequence of *The Woman* ambiguously shows The Woman in the woods, possibly having been raised by wolves, since we see a baby with a wolf carefully licking blood from her fingers. We also see her stealing, even murderously stealing, the den of a wolf when she needs somewhere to 'hole up' and nurse her injured flank. This mouth is not afraid of wolves. This mouth is not afraid of the unclean and improper. This mouth derides the sign of its own restraint.

Figure 3.1 Bite (*The Woman*, 2013, Dir. Lucky Mckee, USA: Moderncinè).

Abjection's creed

To turn to what remains Julia Kristeva's most well-known work in the Anglophone world, *Powers of Horror*, is to find the symptomatic repetition of what Jacques Derrida has called the question of the animal. That is to say, 'the animal' is not her subject in the sense that it so self-consciously is for Derrida: it is not her subject, but rather it is her ground. The subject that she develops is psychoanalysis: 'the animal' is a concept that she repeats. The following long citation is from the opening pages:

> The abject confronts us, on the one hand, with those fragile states where man strays on the territories of *animal*. Thus, by way of abjection, primitive societies have marked out a precise area of their culture in order to remove it from the threatening world of animals or animalism, which were imagined as representatives of sex and murder.
>
> The abject confronts us, on the other hand, and this time within our personal archaeology, with our earliest attempts to release the hold of *maternal* entity even before existing outside of her, thanks to the autonomy of language. It is a violent, clumsy breaking away, 'with the constant risk of falling back under the sway of a power as securing as it is stifling'.[1]

This citation now surfaces in the narrow field of investigation that has sought a conversation between psychoanalysis and animal studies. If we turn to a text written almost thirty years later, we find that 'signification is what determines the human species'.[2] To put it bluntly, this is an index of Kristeva's commitment to maintaining human exceptionalism as necessity in the post-Lacanian theory of subjectivity, to which she has significantly contributed. It is to render rather more forcefully Kelly Oliver's argument in her book *Animal Lessons*, an argument that revisited a range of scholars in light of Derrida's work to further flesh out the ways in which their works police the conceptual divide between the human and the animal.[3] What is particularly salient about Oliver's substantial intervention in the still burgeoning interest in the resources of continental philosophy for animal studies is her demonstration that women scholars have also repeated this division.[4] The scholars in question for Oliver are Simone de Beauvoir and Kristeva, well respected as theorists of feminism and the feminine respectively. They are both also – we should note as Oliver does not – part of the dialectical tradition.[5] Oliver is keenly observant of the problematic ways in which sexual reproduction negatively understood as the metonymy of animality muddies the prospective clean transcendence of woman to the political sphere in Beauvoir (a trajectory that man nevertheless assumes that he can achieve). The seemingly hopeless abject taint of the latter's phrasing – 'the species gnaws at [woman's] vitals' – prefigures Oliver's revision of her extensive extant work on Kristeva, now to bring out the connected issue of the feminine and the animal.[6] On this connection, the long quotation above is instructive. The major invocations of abjection in the art world through major exhibitions in New York and in London in the early 1990s may have reduced it to a fear of border crossings without regard for the anxieties prompting such fear.[7] Closer readers of Kristevan abjection in the humanities – prior to Oliver's book – would certainly have noticed the second aspect of the confrontation with the abject – that which pertains to what she names our 'personal archaeology' or our subjectivity as the attempt to 'release the ["stifling"] hold of maternal entity' by means of the 'autonomy of language'. What Oliver's work incites us to re-read is the division presented here between the subject's relation to the maternal *and* the social relation to the 'threatening world of animals'. As with her tracing of the negatively troped animal figures that 'dog' the emancipatory desire for the transcendence of the 'second sex' in Beauvoir, Oliver causes us to notice that the fragile state of abjection is an animal 'territory' onto which 'man strays'. In other words it is not the proper state or estate of man (even if this estate is always the fictional property of, as Derrida would say, 'the subject that calls himself man'). While an abject anxiety might be mobilized in myriad subsequent contexts of border collapse, including that of the fictions of sexual identity or the nation state (and the phrasing – *border collapse* – is also to pass judgement on what should constitute a border), Kristeva names the foundational forces of abjection as double: the maternal and the animal.

Kristeva's 2010 article titled 'The Impudence of Uttering' confirms the enduring conservatism of psychoanalysis vis-à-vis 'the animal' when she writes that 'signification is what determines the human species'.[8] This article came to attention when searching for any evidence of any impact of the proliferating continental embrace of the animal question in her work.[9] There, Kristeva purposefully invokes the union of language, culture and polymorphous perversity linked to creativity as that of the 'human species'.¹⁰ She thinks this movement through the poetic work of Colette and Proust (and this is seductive, not least because the more mature Kristeva now writes on 'feminine genius' and remarks on her own insistence that women are capable of sublimation in distinction from Freudian orthodoxy).¹¹ But readers should remain alert to her maintenance of the succession from nature to culture (from 'desire-pleasure' to '*jouissance*') as that which humans alone can traverse. This is underlined when Kristeva briefly remarks upon the 'nuptial dance without words' of cats as 'a displaced image of the envied parental coitus'.¹² That this striking image has a feline impetus makes it all the more noticeable that she bypasses works such as that of Derrida. The 'animal sensoriality' of cats is wholly given over to the image of parental coitus: accessing the pleasure of cats is a veiled access to the primal scene. In this world, cats are purely vehicles for human desires (desires that they clothe in a palatable fashion). This continued conservatism should be noted given that some readers misread *Animal Lessons* as simply bringing Kristeva into the fold of animal studies; for example Barbara Creed's essay 'Animals, Art, Abjection' included in a contemporary collection of new writing on *Powers of Horror*, as well as some of Oliver's own subsequent writing that rather softens the critique that she herself made possible.¹³ Christopher Powici, in contrast, suggested that the infrequency of psychoanalytic resources within the broadest realm of ecocritical scholarship is consequential upon Freud's remark, in 'Civilization and Its Discontents', that 'the extermination of "wild and dangerous animals," and the extensive breeding of "tamed and domesticated ones," is a defining characteristic of a highly civilised society'.¹⁴ Powici goes on to caution that 'psychoanalysis may be seen as complicit with the very ideologies and practices that ecocriticism challenges'.¹⁵

Raising animals

This chapter raises animals as our subject, as a subject that is not reducible to a symptom, that is not reducible to a figure – although animals may *also* figure as symptoms and as figures (and this is to recall Derrida's reminder that there is no sure way for readers to know whether his real encounter with the little cat is not *also* an allegorical invocation of *Alice in Wonderland*).¹⁶

We need to raise animals to our attention now, in our time – the time of the sixth mass extinction – when they are disappearing at an ever-increasing and ever-more systemic rate.¹⁷ However, to raise something to attention also speaks to an anthropometric scale and sensorial form of attention as it reveals something before us, or stands it before us, presented for our visual attention. The vertical and the visual come together in the way that we typically imagine human evolution: we are all too familiar with the ascent of man, diagrammed as the vertical ascension of a white man walking into the future, away from his more horizontal, hairy, dark, hominid ancestors. That diagram finds its psychoanalytic support in Freud's well-known footnotes to 'Civilization and Its Discontents', in which he writes that the 'dimunition of the olfactory stimuli seems itself to be a consequence of man's raising himself from the ground, of his assumption of an upright gait'.¹⁸ These footnotes suggest that to the extent man becomes a standing animal, an erect animal, he becomes one that visually surveys his domain rather than smells a terrain. And on this terrain, humans are those who raise animals, who practice animal husbandry in the particular agricultural sense as well as – and this is not necessarily fundamentally separate – in a philosophical sense of the management of concepts. Impressively unrelenting, this movement of elevation ascribed to the human is ingrained within the form of dialectical procedure as an overcoming, *Aufheben*, and in French, *relever* – to raise up.

In raising animals, I want to augment Oliver's caution regarding the double abjection of the animal and the feminine in Kristeva through the prism of Lucky Mckee's 2011 horror film *The Woman*, a film that offers an acute if troubling meditation on these concepts.¹⁹ While horror might sound like an unlikely place in which to meditate on any concept, the genre has frequently invited just such philosophical work (and not least in questioning the morality of the family in such classics as *Texas Chainsaw Massacre*).²⁰ Indeed, in his own excursion into the genre, Cary Wolfe remarks on horror film as the 'mass cultural heir of the Greek tragic drama'.²¹ Barbara Creed's classic work of feminist film theory, *The Monstrous Feminine*, from 1993 notably invoked horror with the work of Kristeva as that which offered the possibility of the feminine as a power before the law of the father.²² Drawing on rape revenge and cannibal horror genres, *The Woman* viscerally opposes the patriarchal nuclear family as it both exposes the banal structure of domestic violence that holds that family in place and embodies the psychic terrors drawn through sex and species difference on which it feeds.[23] That the film draws on genres liable to entrench the misogyny on which they trade allows, in this instance, a substantial departure from that outcome not merely in the denouement but with sympathy for the titular Woman throughout. To some it might seem heavy-handed to cast the male lead in *The Woman* as both brutally patriarchal and a lawyer – to position one as the expression of the other. Others might feel interpellated by that very union.[24] Yet in our times when 'monsters' do indeed walk the earth, when the so-called Leader of the Free World did not have his campaign for that position derailed by the film exposing his crude discussion of the sexual benefits of fame, a discussion infamously centring on his boast that he could 'Grab 'em by the pussy … do anything',[25] and, amongst numerous other horrors, he went on to appoint a man credibly accused of sexual assault to the Supreme Court of the United States: in these times I love *The Woman*.[26]

The Woman further allows for reflection upon the canonical feminist theorization of gender of the 1970s and 1980s and its critical appropriation of Claude Levi-Strauss's anthropological equation of culture with the exchange of women between men.[27] I will thus take seriously Kristeva's own admission that her reflection upon the feminine will offer 'no solace' (since, as she particularly emphasizes, it structurally cannot offer an alternative representation – or alternative *to* representation).[28] This is absolutely not about culturally constructed notions of what is 'feminine' and what is 'masculine'. I will come back to this crucial structure of exchange in particular, given that Kristeva upholds it as necessity alongside her otherwise notable amendments to Freud. The consolidation of units of exchange points also to the definite article that the films very title announces – *The Woman*. Technically, there are several women in *The Woman* (Belle, the mother (Angela Bettis); Peggy, the pregnant teenager (Lauren Ashley Carter); Darlin', the little girl (Shyla Molhusen); Genevieve, the teacher (Carlee Baker) and the eponymous Woman (Pollyanna McIntosh)).[29] While cinematic attention is particularly given to the feral woman captured in the woods by the film's leading man – who chains her in the cellar, ostensibly framed as his family's civilization project, philosophical attention is arguably solicited with regard to the absolute quality or even archaic type that the '*the*' instils. Given the schematic cast and locations, we are also encouraged to see the country house in which the film is set as The House, the family as The

Family and perhaps to pursue a spatial analysis of metaphysics as a gendered *oikos* or home economy.[30] Yet, as I will argue, what sediments these categories is the implication that they are distilled in relation to the concept of The Animal.

'Have you fed the dogs yet?' This anxious question is frequently repeated in *The Woman*. Along with the agitated demanding sound of barking, somewhere, we are given to understand, there are hungry dogs, close to but never inside the house. They do not patrol the grounds of the house. They are not taken for walks. They do not accompany the father-lawyer, Cleek (Sean Bridgers), on his military-grade weaponry-equipped nocturnal hunting trips. The dogs never leave what I'm referring to as the kennel out of convention, but more descriptively it is a caged part of a large shed. Thus anything but the most rudimentary signs of the domestication of dogs are scarce. They are not the 'humanized animals' that we call pets with the grid of species that Wolfe suggests as the provisional means by which to account for the various typologies orchestrated by *The Silence of the Lambs*.[31] They are not on guard: but has anyone fed them? If the category of companion species can be described as a mode of relating that situates those who eat with others (company describes those who gather around bread, '*cum panis*', as Donna Haraway suggests), that companionship has been subjected to very tight controls.[32] While everyone in this family knows about the dogs, refers to them as 'the dogs' without invoking any individual names and the daughter Peggy volunteers to feed them, we only ever see Cleek and his son doing so (and with a cruelty reducible to a strict law of exchange: accept food or get beaten).[33] The shots that show this enclosure – until the closing sequence – remain shy of who or what is being fed, beyond two leashed German Shepherds, in the 'kennel within the kennel' at the rear of the enclosure. The sound of barking is not contained to the shed but can be heard all around the property, including in the cellar in which the Woman is restrained, pairing the liminal spaces of kennel and cellar. Both spaces belong to the house but are peripheral to it; both come with locks and chains and weaponry. The force with which these spaces are fortified contrasts with the depleted maternal figure within the house itself in which Belle flinches at the mere sound of Cleek's raised voice. The kennel and cellar may be secretive spaces, but all family members are psychically bound to them in various relations of fear and fascination. Unlike the inhabitants of the kennel, all family members are shown the Woman *and* we are shown their various reactions to her, her condition and their own role in maintaining it. When Cleek takes his family down into the cellar to see their 'civilization' project – the Woman – he warns them not to get too close because she 'likes to bite' and shows them his severed finger. 'We can't have people going around the woods thinking they're animals', he says, 'it isn't right, its not safe'. While Darlin' reacts curiously to this threat of devouring by both bending her finger as if she too had lost part of it *and* smacking her lips as if she were the one biting a finger off and swallowing it, Cleek explains that this is a secret project that they would all share in looking after and keeping hidden, adding that 'its just like the dogs'.[34] This name – *the dogs* – houses secrets.

Secreted away from the public, the family members know about 'the dogs'. They are raising them after all, even as this familiar name fetishistically grasps what the kennel actually houses. (The fetish works in the classical manner of disavowal structured as 'I know, but I don't know', naming what is foregrounded in the kennel but maintaining the cover-up of what unnamable difference lies to the rear.) Relatively restrained overall as a horror film, *The Woman* saves its biggest shock – in line with genre expectations of mounting horror and escalating violence – for the penultimate revelation for the audience of just what has been going on with the dogs for all this time. 'All this time' might index the duration of the film, but perhaps a wider conception of

Figure 3.2 Introduction (*The Woman*, 2013, Dir. Lucky Mckee, USA).

time as such, with the dogs taking on a phantasmatic sense of a perpetually ravenous beast that must be fed to be contained but whose hunger can never be sated, whose jaws are always open. The peripheral spaces themselves – kennel and cellar – metonymize dangerous mouths. If an anxiety-laden duty surrounds the kennels – a duty that has nevertheless become quotidian – fear and fascination envelop the cellar and the captive that it restrains. Belle, the mother, is fascinated with her strength, almost on her side, but fatally fails to help the Woman when she has the chance (being unable to overcome her terror of Cleek). Peggy is aghast at the situation and brave enough to at least intervene when the Woman is being brutally washed with a power hose, and she is the one who will finally release the Woman when there is no other force for help possible. Like his father, Brian (Zach Rand) evidently understands that this ostensible project of 'civilization' would not exclude her sexual abuse, immediately asking, 'do we really get to keep her?' The civilized woman is a kept woman, in this conceptual history. Later Brian takes sadistic pleasure in her captivity, spying on his father's late-night visit and rape of the Woman, and subsequently returning to inspect her to the point of torture himself. It is Peggy that puts a stop to his cruel investigation of her breasts. Meanwhile, Darlin' places her new toy radio outside the cellar door so that the captive might enjoy some music, like she does.

A/basement

'Civilization and Its Discontents' is not only the tale of the erection of man; it is also that of the hostility of women. Freud writes: 'women soon come into opposition to civilization and display their retarding and restraining influence ... the woman finds herself forced into the background by the claims of civilization and she adopts a hostile attitude towards it.'[35] The reversion of 'women' to 'the woman' is ironically apposite, given Cleek's explicit framing of the Woman as the family 'civilization project'. In this context it refers to the task of civilization leading men – those he deems capable of sublimation – away from the family (the law of which their fraternity nevertheless shepherds). Sublimating his sexual drive into the work of civilization 'estranges' the man from the home. It is not even that women are themselves simply more interested in family and sexual life than what culture might avail. Rather, in Freud's phrasing, they '*represent* the interests of the family and sexual life'.³⁶ As such they are cast as signs, always and already.

Caught in the woods, the Woman is covered in filth and, inextricable from the ostensible 'civilization project', she must be made clean. Cleek's manifest narrative identifies species boundaries as the prime offence in need of correction: he names her as 'going around the woods' mistakenly 'thinking that she is an animal', implicitly requiring rescue. He also identifies her as stinking – and we should recall that part and parcel of the becoming-erect of the human subject, for Freud, lies in the repression of smell – the repression of the proximity of filthy, animal, bodily odours.³⁷ But in so far as a white actress is effaced by dark dirt and chained in a cellar, the film risks a spectral figure of racial difference in the Western, and in particular, the North American, imaginary. That imaginary violently organizes race within a hierarchy given literal and symbolic freight in the historical practice and continuing legacy of slavery. This figure is at one with the European history of soap as that which sold the transition from dirty to clean as coterminous with one from black to white.³⁸ While Kristeva suggests that '[t]he body must bear no trace of its debt to nature: it must be clean and proper in order to be fully symbolic', the force field of the concept of nature draws the black, along with the woman and the animal, into its static holding pen.³⁹ Attempting to redraw the edges of the Woman, without any abject 'bleed' of categories, firstly by cleaning her – and it is instructive that this apparently simple task is always in excess of its ostensibly basic remit – does not raise her to the status of 'fully symbolic' (as her unclean bloodied face in the film's conclusion will announce). When the Woman is cleaned, Cleek enlists his family making them complicit with the task. It viciously encompasses the power hose (applied by Cleek himself) followed by water that is close to the point of scalding (applied by Belle): both approach a skin-stripping level of cruelty. If the removal of the blackening filth results in her being rendered more properly white, the Woman's elevation in status is yet minimal. She remains a captive that Cleek's family are expected to accept in all the abasement that such captivity proposes.

In the shade of the Enlightenment, and somewhat infamously albeit within parentheses, Freud casts the sexuality of women in racist tones, by means of both denigration and colonial exploration:

We know less about the sexual life of little girls than of boys. But we need not feel ashamed of this distinction; after all, the sexual life of adult women is a 'dark continent' for psychology. But we have learnt that girls feel deeply their lack of a sexual organ that is equal in value to the male one; they regard themselves on that account as inferior, and this 'envy for the penis' is the origin of a whole number of characteristic feminine reactions.[40]

Responding to Freud in her book exploring the colonial imaginary within psychoanalysis – a book that took *Dark Continent* for its very title – Ranjanna Khanna notes the transference that Freud himself performs: transferring the 'shame' that '*we* need not feel' onto the girl and the woman. They are the ones more properly shameful:

> Perhaps fearing her difference, he makes her other, obliterating the specificity and difference of her body by turning it into a fetishized metaphor of the unknown: 'dark continent,' and it is defined as lack ... Although the Other [in Freud] is not intrinsically racialised or sexualised, it does seem that travel and exploration are the instigators of a theory of the Other.[41]

Khanna observes that Freud magnifies the colonial, mysterious 'aura' of the 'dark continent' through his retention of the term in English. The term, as she notes, first 'came into use in H. M. Stanley's explorer's narrative about Africa: *Through the Dark Continent*' – a narrative laden with that author's own anxieties about women. 'The 'dark continent', she continues, 'connotes a great deal, but denotes nothing: it is indefinable, and it is primitive, but it allows its explorers a heroic narrative of discovery and a feminisation of the land'.[42] The Woman, we recall, was captured in/rescued from the woods, a terrain that we only otherwise see Cleek explore.

The abasement of the Woman delivers a shock both to Belle and Peggy and to the 'sex/gender system' that became second nature to second-wave feminist thought.[43] Gayle Rubin's canonical essay from 1975, 'The Traffic in Women', first articulated this system through appropriating the language of exchange and that of signification drawn from Claude Levi-Strauss and Jacques Lacan in order to foster not a prescription for what is called gender (as necessity) but a description of a particular system (as contingency).[44] The political hope was thus that a 'revolution in kinship' might prompt a wholly other system.[45] In so doing this would further dismantle the compulsory heterosexuality also and not so surreptitiously articulated by this exchange of signs.

Asked to give an account of the term 'gender' as an entry in a Marxist Dictionary by a group of German feminists in 1983, Donna Haraway felt that Rubin's 'sex/gender' system was so sedimented in Anglo-American feminist theory for this to be her expected theoretical anchor. Given the authoritative stature of a potential work of reference and that her task was rendered yet more complicated by the editors proposed translation of this Dictionary into not 'merely' four European languages (English, French, Spanish and German) but also Russian and Chinese, a task of translation that itself begged the question of the universality of all the various discourses leading to Rubin's usage, Haraway produced what we might call a critical genealogy of the term instead. Its current and deceptively simple usage not withstanding, even – perhaps especially – the German term '*Geschlecht*' drew on 'sex, stock, race, and family'.[46]

While noting that Haraway's genealogy is not 'post-gender' in the misconstrued temporal sense of 'post-feminist' but most certainly is *post* a dialectical economy of sex and gender (or nature and culture), we should pause on her attention to the acute historical discrepancy in the formation of symbols versus objects of exchange upon which black feminists such as Hortense Spillers and Hazel Carby insisted.[47] In the development of the 'sex/gender system' as a central focus of feminist analysis, white feminists did not see the glaring difference in which they were implicated, summarized by Haraway thus: 'Free women in US white patriarchy were exchanged in a system that oppressed them, but white women *inherited* black women and men.'[48] In this divided system, as Carby underlines, it follows that white women gave birth to humans within a symbolic economy, but 'black women gave birth to property ... to capital itself in the form of slaves'.[49] Haraway details the consequences of this structural difference:

> Slave mothers could not transmit a name; they could not be wives; they were outside the system of marriage exchange. Slaves were unpositioned, unfixed, in a system of names; they were, specifically, unlocated and so disposable. In these discursive frames, white women were not legally or symbolically human; slaves were not legally or symbolically human at all.[50]

Abruptly faced with the filth-covered and shackled Woman in the cellar, one not even given a name, and one clearly at the disposal of both father and son, Belle and Peggy are faced with an unspeakable crisis in the categories of exchange, sign and property in which they are enmeshed.⁵¹ As tokens of exchange themselves, they are yet called on to maintain the actual captivity of one whose category falls outside legal redress, even as the titular definite article promises some kind of restoration (and that this expectation is conceivably finessed by the fact that McIntosh is a white actress and thus the confrontation with the historical divide between black and white women is not more fully acknowledged). Kept underground in this liminal space that itself might be thought of as 'unclean' or even 'unconscious' compared to the interior or 'conscious' space of the house 'itself', the Woman is kept in the dark. Yet the crisis of category, and of ethics that those categories claim to license, is brought to light in ways that cannot but recall the structure of the uncanny. This uncanny return is prompted not least since they have already been participant in keeping one of their own kind in captivity – as we will come to learn. Significantly, in the case of this other prisoner, this captivity has been without promise of the so-called civilization. The latter crisis in kind does recall the attempt to secure species by means of the grid of which Wolfe wrote, this time in reference to that of 'animalized humans' (to which the last section of this chapter will return).⁵²

La Bête et le Souverain

Above or beyond the law, alternatively in violation of it, the political figures of the beast and the sovereign seem to form two distinct poles while sharing what Derrida calls a 'troubling resemblance', this 'being-outside-the-law', this 'reciprocal haunting'.⁵³ Indeed this uncanny proximity cannot but be heard in the French title of Derrida's seminars, *La bête et le souverain*, where the difference between *est* and *et* ('is' and 'and') is inaudible. The potential substitutability of the beast for the sovereign is a murderous substitution: where 'man is a wolf to man', he is a lethal threat. Such is the thicket of animal figures especially wolves or The Wolf in political discourse that Derrida can weave between Rousseau, Freud and Little Red Riding Hood in a matter of lines. The 'recurrence of the lexicon of devourment' is common to all: political theory, psychoanalysis, fairytales.⁵⁴ Derrida cautions us not to 'forget the she-wolf'; for example she that suckled the twins Romulus and Remus at the foundation of Rome.[55] His French pauses on the gendered inflections of *La* and *Le*: the female beast and the masculine sovereign – 'so *what*? So *who*?' – repeating this inflection frequently across the seminars, although perhaps without enough development (enough to pique the interest of feminist readers; not enough for the majority of his readers to make it unavoidable).[56] Derrida's use of 'what' and 'who' here recalls the much earlier interview 'Eating Well', in which he first explicitly addressed the 'question of the animal' and the death sentence it ultimately commands between those designated a 'who', or a 'thou' and a 'thou' who 'shalt not be killed', while the implied 'what' is pushed outside of ethical consideration and may be put to death without criminal offence (as elaborated in the previous chapter). Just as 'the animal' was diagnosed as the *conceptual* corral that we must undo, 'the beast' in his seminars does not engage with living animals in any detail but does refer us to the consequences of our bestial imaginary for the living. Kristeva's abjection of the animal and the woman sits awkwardly with Derrida's acknowledgement of the she-wolf, awkward because of its resistance to representation.

This mouth must be cleaned up, must be taught to eat properly, to eat cooked food and to say the words 'thank you' (the enunciation of the latter as a form of contractual agreement produces a scene which, of all her degradations at the hands of Cleek, is singularly unbearable).[57] This mouth must be taught to exchange food for words in acceptance of the dominant means of signification. Cleaning up the cellar under Cleek's instruction, the family follow that instruction but with varying degrees of compliance and hesitation regarding the oddness of this sudden order. Only Darlin' is excited: perhaps there are mice and she should fetch some cheese. Belle makes cookies for her children: on request they are gingerbread men. 'Would the animal lady like to eat a little man?' Darlin' asks her, generously: her thought of a gift contracts into a classical Freudian exchange of signs that condense oral and genital, cannibalism, identification and impregnation. Darlin's own identification here aligns her with the power to impregnate by means of the companion gingerbread men (she wants to give her a baby), but Belle blocks contact.[58] Affirming the Law and the straight lines of identification in Oedipal orthodoxy, she dismisses Darlin's wish, saying that it is her father who is euphemistically 'helping' the Woman.

While the core image of *Powers of Horror* that sticks in the academic imagination might be Kristeva's opening reference to the skin on the surface of milk inducing the urge to vomit, the book's third chapter on 'Filth and Defilement' obliges us to return to the cultural primal scenes of the Freud of 'Totem and Taboo'.[59] That text, as indicated in the previous chapter, vacillates on whether our ancestors were animals or the Father, incrementally crafting the way to installing that Father at the origin with animals only ever serving as His substitutes. 'Filth and Defilement' also makes the encounter with the feminine and the animal unavoidable. At the very end of his text, Freud consolidates the beginning that 'Totem and Taboo' dramatizes; quoting Goethe, he writes, 'in the beginning was the Deed'.[60] Kristeva observes that by this ending, one could be forgiven for forgetting that *two* crimes announce the origins of culture in Freudian legend: the murder and cannibalism of the father *and* incest with the mother. The latter 'disappears' even though the attempt to explain the severity of incest dread had driven Freud's investigation into the paucity of anthropological accounts of culture and even though it is the apparent motivation for the foundational patricide.[61] Immediately, Kristeva is clear that she *does not dispute* Freud's account of this patricide and its privileged setting as the 'keystone to the desire henceforth known as Oedipal'.[62] This is crucial to note since Oliver sometimes gives the impression that Kristeva's attention to the mother diminishes the investment in the father that Freud consecrates, rather than strictly redresses his lack of attention to this other structure. Rather, Kristeva expands upon what she considers to be the 'two-sided formation' of the sacred ('sacred' given that this formation of law is bound up with that of religion). These two sides, however, are radically different in kind and in consequence. There is:

> One aspect founded by murder and the social bond made up of murder's guilt-ridden, atonement, with all the projective mechanisms and obsessive rituals that accompany it; and another aspect, like a lining, more secret still and invisible, *nonrepresentable*, oriented toward those uncertain spaces of unstable identity, toward the fragility – both threatening and fusional – of the archaic dyad, toward the non-separation of subject/object, on which language has no hold but one woven of fright and repulsion?[63]

Both sides hold negative affects, but as she continues, '[o]ne aspect is defensive and socializing, the other shows fear and indifferentiation.'[64] Paternal murder founds the law, the law that begins with a 'thou shalt not kill', and hence the possibility of a social world, defending *against* murderous lawlessness: maternal incest threatens a relapse. What is both a temptation and a problem for feminist readers is that while we may desire the end of the law as authorized by the Father, this framework holds that this other aspect cannot offer a replacement: it is nonrepresentable. It is important to be clear about the structure that Kristeva nevertheless upholds. She writes:

> If the murder of the father is that historical event constituting the social code as such, that is symbolic exchange and the exchange of women, its equivalent on the level of the subjective history of each individual is therefore the advent of language, which breaks with perviousness if not with the chaos that precedes it and sets up denomination as an exchange of linguistic signs.[65]

There is thus a distinction between the category of 'women' *qua* tokens of exchange (a sign amongst other signs given order by the paternal signifier) and the feminine that can produce no sign of its own. Moreover, this invocation of the social code 'as such' as the exchange of women between men inextricable from the advent of language, and hence the exchange of linguistic signs, speaks not only to the Freud that Kristeva more frontally addresses but also to the structural anthropology of Lévi-Strauss.[66] Lévi-Strauss even wrote, 'women themselves are treated as signs, which are misused when not put to the use reserved to signs, which is to be communicated.'[67] Lacan countersigned this sexual contract, agreeing that the 'communication' of women between set groups of men would '*guarantee* that the voyage on which wives and goods are embarked will bring back to their point of departure in a never-failing cycle other women and other goods, all carrying an identical identity'.[68] In Kristeva's psychoanalytic lineage, if there are to be signs, if there is to be signification as such, replete with transcendental signifier secured by the name of the father in the train of thought sedimented by Freud's totemism and consecrated by Lacan, then the equivalence of women and words must be endorsed. Unlike Rubin, Kristeva does not counter Lévi-Strauss.

The long citation previously discussed – from the opening pages of *Powers of Horror* – sketched two axes: the social abjection of animals and the subject's abjection of the maternal. Yet at the point where Kristeva introduces the 'two-sided sacred' with the caveat that 'What we designate as "feminine," far from being a primeval essence, will be seen as an "other" without a name', animals have sunk below even this non-status seemingly not able to be raised even as nameless.[69] The two powers that Kristeva directly invokes, powers that 'attempt[ed] to share out society', are masculine and feminine. 'One of them', she says, is 'apparently victorious' but through its very 'relentlessness against the other' reveals that it is 'threatened by an asymmetrical, irrational, wily, uncontrollable power'.[70] The symbolic is not strong enough to 'dam up the abject or demoniacal potential of the feminine. The latter, precisely on account of its power does not succeed in differentiating itself as *other* but threatens one's *own and clean self*. It is a locution that appeals to us, calls out to us, speaks almost in the manner of leftist appropriations of Hegel's master-slave dialectic in which the slave, or bondsman, whose very work – compelled though it might be – promises to eventually enable him to rise up in triumph over the master, or lord, whose might is ironically belied by his non-labouring body.[71] For Kristeva, however, this asymmetrical, irrational, wily, uncontrollable power cannot absolutely overcome the one whose name and whose economy of exchange 'is defensive and socialising', despite being 'no less virulent'.[72] To the Anglophone ear, 'wiliness' perhaps conjures up a seductive animal style, the cleverness of an animal primed to outsmart rivals, threats or prey. Thus while the noun 'animal' does not appear in the section on the 'two powers', the phrasing perhaps bears its figural trace. However, where the concept of 'the animal' is at work, that is to say in metaphysics, in dialectics, animals will struggle to exceed that figural implication. Indeed the battle conducted between master and slave in Hegel's dialectic is precisely over 'nothing' but 'prestige', and as such it elevates the master who is willing to sacrifice his life over nothing above the animal body of the slave.[73] While it is the frontal argument of *Totem and Taboo* to position patricide as the origin of culture and/or as the Law, it is Derrida who pursues the logical implications of this for non-human animals. Not only can animals never rise to symbolic status within the Law themselves but they can never commit an 'infraction of the law': animals are never criminals.[74] Just as Derrida has asked what would happen to fraternity should a sister or an animal enter it, we might also ask what would happen to psychoanalysis if we were to follow ethological curiosity and allow that non-human animals may produce other laws, not necessarily commensurate with human law? Will our Case fall when it no longer sits in place as the Case?

What if the choice is not in the either/or form *between* a matter of raising all abjected others to the same status – a politics of representation, *or* of locating a pure force of resistance outside of representation, but that the field of our encounters demands other architectures?

Where there is food, there is also a polluting object. At the 'boundary between nature and culture, between the human and the non-human', Kristeva tells us: 'all food is liable to defile'.[75] Cleaning up means dealing with the dirt and the 'frailty of symbolic order'.[76] The sudden organization of the cellar is absolutely under Cleek's direction, from cleaning the space to cleaning up after the Woman; this undertaking in 'civilization' raises the game of both 'mere' housework and maternally identified toilet training. Brian is seconded to both of these dirty and feminized tasks. He is quietly if resentfully compliant in the cellar: in the kennel, feeding the unseen inhabitant of the cage within the cage he accidentally puts his hand in her shit. In resentment and disgust, he calls her a 'bitch'.[77]

Feeding the dogs and cleaning up after this 'bitch' in the enclosed, sectioned-off spaces, Brian is yet contaminated by shit. Legible as the metonymy of the attempt to control corporeal orifices, these spaces will never be clean. Resentful of his delegation to cleaning duties and especially of the accident that sullies him with shit, in echo of his own sphincteral training, Brian returns to the breast. He returns not to nourish himself on food but to exact punishment upon the Woman, refusing the breast as that which provides milk and was mastered by the signifying substitutions of language. Brian obliges the breast to surrender blood.

Mouthing off

We are primed to expect the defensive structures that architecturally and psychically map out the space of the house and its abjected out-buildings, the cellar and the kennel, to be breached. We are primed to expect this breaching to emerge from the interior – to be a breaking out more than a breaking in. It is a horror film – bolting a door will never bolt the danger out but always lock it in. Narrative expectation has focused upon the Woman, and while she is on one level a stranger, her 'naming' in the form of the definite article invites us to expect and to desire an eruption of the feminine within this patriarchal fortification. Perhaps she will even ally with 'the dogs', whose agitated barking she has heard.[78] While that desire is indeed given dramatic expression, a 'sucker punch' comes from a surprise source.

'Anopthalmia'. The medical term is so unfamiliar, and the eruption of violence that accompanies its utterance is so 'eye-catching' that it is easy to miss. Belle prompts its first usage when she loses her temper at long last with Cleek: she is completely incensed by his refusal to agree that Brian has done any wrong and deserves any punishment. In the ensuing argument Belle defends the rights of the Woman specifically as those of a 'human being' while suddenly making reference to what is 'going on with the goddamn dogs' as 'enough to put [Cleek] in prison'. He responds, highly cryptically for the cinematic audience, by simply saying the word 'Anopthalmia', followed directly by '*Your* shame'. The Greek term has yet to make sense in the narrative, but retrospectively, the metonymic chain is made to cluster around *her shame*: the dogs; congenital blindness; the mother. If there is a threatening secret, it is all her fault, he says. As we will see, that shame is anchored in a fault.

In Kristeva we find a distinction within her understanding of language: the distinction between *mapping* and *laws*. She speaks of a 'primal mapping' that

> shapes the body into a *territory* having areas, orifices, points and lines, surfaces and hollows, where the archaic power of mastery and neglect, of the differentiation of proper-clean and improper-dirty, possible and impossible, is impressed and exerted.[79]

This arises from 'maternal authority', and it 'is the trustee of the *mapping* of the self's clean and proper body; it is distinguished from paternal *laws* within which, with the phallic phase and acquisition of language, the destiny of man will take shape'.[80] In Kristevan psychoanalysis these impressions map territory but do not write the law. The law distinguishes and organizes discrete elements – including paradigmatically as we have heard, women and words – and represses any maternal authority that would take matters into its own hands. Rites attempt to manage the border: the border between the semiotic of the body and the word of the law: from rite into right. Not simply 'filth' but 'defilement' comes to be what Kristeva calls the 'translinguistic spoor' of our 'most archaic' boundaries with the mother, the abjection of which is given salve through ritual.[81] Again animality leaves a figural trace or 'spoor' in her prose:

animality is not her subject. With perhaps an ambivalence towards both the scholar on whom she has published extensively and also towards the animals from whom she would learn, Oliver notes that it is 'the animal that puts the teeth into her notion of the abject-devouring-mother', holding on to this bite even as she knows that what puts bite into this animal is a figure of carnivory.[82] This bite, that always exceeds need, returns us to the Woman's initial response to her captivity: to refuse to circulate the sign of her own conscription.

In a film as primal as this, and with the severance of Cleek's finger swallowed as inconsequential flesh and his ring rejected early in its narrative, we expect that a carnivorous appetite will resume. In that troubling resemblance of beast and sovereign, Derrida reminds us that both stand outside or above the law: ruling the household Cleek feels utterly immune from incrimination, up to and including murder. Thus, he cannot tolerate being told that he '*can't*' do anything and repeats the word in rising tones as he strikes Belle to the floor. Domestic violence is orchestrated precisely around Cleek's absolute intolerance of not 'being able', the precise quality that, as Derrida notes, the virile figure of carno-phallogocentrism claims to possess ('man' is always the one who *can* speak, *can* reason, *can* mourn, etc.).[83] In the very next scene a second woman is struck down. Cleek views the spontaneous home visit conducted by Genevieve, the schoolteacher (at which she raises concerns regarding Peggy's pregnancy), as an affront to his authority. This affront is all the worse since it is made 'within his own home', as he remarks. This is a home in which women do not step out of place, and in which there can be only one 'Teacher'. The word '*can't*' and the flare of anger it triggers repeat when Peggy tries to stop Cleek's subsequent assault upon Genevieve. Holding Peggy up by the scruff of her neck, thus making a mockery of being able to stand upright, Cleek again scathingly repeats, '*I can't?*' as the camera makes a 360-degree pan around them (a still rare shot connoting mastery of all that can be seen since the enabling cinematic apparatus remains hidden, albeit one now finessed by digital rather than cumbersome analogue technology). From here the film's rapidly shifting final sequence of punishment and retribution, all of which involve forms of cannibalism, is given full rein.

The dogs are really barking now, and their agitation is heard across the film's cuts between Cleek dragging the schoolteacher towards the kennel, Belle coming around in the kitchen and the Woman pulling at her restraints in the cellar. In the montage of reactions to the rising sense of panic, Darlin' asks, 'Mama, what's happening to the doggies?' She is the only one to ever refer to the dogs with affection. The film immediately cuts to the terror of the teacher about to be fed to them, allowing for the persistent impression that these are 'only' brutalized dogs.

The second time we hear what likely is a still an utterly mysterious word, Cleek derisively asks the teacher, now thrown to the dogs – literally and figuratively – whether she can say 'anopthalmia', repeating the word with scornful enunciation. Stressing each syllable, Cleek identifies his powers of elocution with the circulation of discrete units, with signs, with the paternal signifier as that which organizes the social. The word itself identifies the blindspot of sexual difference.[84] 'A study of dreams, phantasies and myths', Freud writes, 'has taught us that anxiety about one's eyes, the fear of going blind, is often enough a substitute for the dread of being castrated'.[85] As revoltingly voyeuristic as he has been throughout, and in counterpoint to anopthalmia, Brian seemingly addresses the dogs saying, 'let's see what you can do.'

On the one hand, the film rapidly cuts between all simultaneous action taking place with each protagonist in such a way as to magnify the tension and the horror (this is a genre that is hardly going to respect the law 'thou shalt not kill'). On the other hand, one of the most psychically challenging matches on action occurs when the Woman is finally freed (by Peggy) and, of the two spaces that Belle could possibly stagger towards, she chooses to go to the cellar drawn by its open door as the apparent cause of the chaos. *The Woman* now reveals the family's other secret. Raised as one of 'the dogs', raised in the same ravenous state, raised entirely to eat, to anticipate live food without regard for what is 'good': the eldest daughter has been kept crouched and snarling at the back of the kennel. This is the locus in which the concept of sexual difference is cinematically concatenated with the concept of the animal. This abject beast-woman initially seems well described according to Wolfe's species grid – here gravitating towards the pole of the 'animalised human'.[86] That said we have already noted that 'the dogs' scarcely occupy any developed domestic role dogs usually inhabit. If 'animalised human' is not descriptor enough to account for the beast-woman revealed in the kennel within the kennel, neither is the 'animalised woman' that Carol Adams introduces when she modifies Wolfe (in a manner coincident with her sociological understanding of sexual politics; i.e. analysis must include 'gender').[87] What has to be grasped is the way that the specific figuration of anopthalmia, while shocking, also makes systematic use of the discourse of disability in tandem with sexual difference construed according to Freudian orthodoxy in order to constrain this creature to the horizontal animal plane.

In its match on action, *The Woman* pairs this obscene instrumentalization of the animal and the feminine as the beast-woman is set upon the teacher with Peggy setting the Woman loose only for her to run straight into Belle. The sheer shock of the condition of the eldest daughter in the kennel is matched with that of the Woman's actions. What is most difficult to stomach is the evisceration each performs as the action at a distance under the instruction or perhaps the wish of another (the instruction of Cleek; the wish of Peggy). The match between these scenes is striking in that it obliges us to see Peggy's actions as deliberately aggressive towards her mother – as if she harnesses the Woman as an attack dog, wishing for her death – a latent possibility significantly at odds with the manifest narrative of desiring to save her teacher and letting loose this unknown force as the last resort.

In terms of the narrative, we can understand the Woman taking revenge on Belle, the woman who failed to help her when she could have made a stand, appalling as this scene is, especially if we want *The Woman* to revolutionize all the women. But in light of both Kristeva and Freud and the identification that ingestion awards, including paradigmatically through cannibalism, it makes sense that when the Woman kills Belle – this diminished mother – she eats her face. She eats her face and takes her place. This sense making, we have to acknowledge, is the work of exchange, of substitution. It is the removal of the mother by the Mother. Countering the non-appearance of anything more than that which is 'like a lining' in Kristeva's frame, Derrida's attention to the lexicon of devourment as that which inheres in the law, rather than that which is expelled as its abjected fear, opens a path to the hungry heart of the law:

> Devourment, vociferation, there, in the figure of the figure, in the face, smack in the mouth, but also in the figure as trope, there's the figure of figure, vociferating devourment or devouring vociferation. The one, vociferation, exteriorizes what is eaten, devoured, or interiorized: the other, conversely or simultaneously, i.e. devourment, interiorizes what is exteriorized or proffered.[88]

While the Woman barely utters any words throughout the film, in contrast to the extraction of 'thank you', she recognizes Peggy's pregnancy with the word, with her vociferation of 'mother'. That is the term that is carried, the one that she circulates.[89]

Suddenly the kennel is no longer a space in which a bloody spectacle can be secured on one side of the cage only nor designated as that over which men have authorial direction. The Woman appears in the door to the kennels with her face returned to being filthy, now not blackened but reddened with consanguineous blood drawing a maternal line.[90] She swiftly fells Brian with a sharp piece of metal. As her teeth were strong enough to sever Cleek's finger earlier in the film, now her bare hand can penetrate his torso and pull out his beating heart. Taking a cursory bite out of this heart, which event, for the sake of the vengeance-driven horror, is visible and mortifyingly comprehensible to the father who dies in and of his punishment, the Woman then lets the now truly heartless father collapse.

If devourment and vociferation give the figure of the figure that is a face to the Woman and brings her, *la bête*, into a substitutive position with both Belle *and* Cleek, another creature remains virtually muzzled. Consistent with 'the dogs', the film gives no diegetic name to the sightless creature that, sniffing the air, follows the Woman out of the kennels on all fours. It is unclear how best to refer to her. The credits and novel refer to her as 'Socket'. She has no eyes, only sockets. The name contracts to the singular: Socket. Perhaps 'Anopthalmia' is her proper name, rooting her in congenital blindness, but 'Socket' is a pet name, a diminutive, suggestive of a sock puppet. The sucker punch is a boxing move that the boxer doesn't see coming because it lands instead of the one that was only a bluff (perhaps he was too busy suckling at the breast, like a baby). Here it turns into a 'socket punch'. This singular wound inflicted upon the eyes summons imagery for which the psychoanalytic mainstay of castration is unavoidable. Fixing the sign of castration to that which cannot be overcome – blindness as congenital fault – works very hard indeed against any conceptual shift.[91] Wolfe once levered Derrida's revaluation of 'being able to suffer' as the weak ability that would confound the old hierarchy of ability versus privation and thereby affirm a new and generalizable condition of the living. In his acute argument this would allow for the '*dis*abled' to become newly inscribed as those that might expose the humanist fantasy of exhaustive vision, and thus 'see' the truth of vision now revealed as necessarily partial.[92] *The Woman*, however, contracts the beast-woman to being strictly *incapable* of sight. Blind, animalized, castrated: she can never stand upright; she will never command moral uprightness or rectitude, never survey a domain but will remain in the stink of terrain. 'Her shame' comes out of the kennels, levelled by fault, by shame, by what Luce Irigaray called the 'blindspot in an old dream of symmetry'. That 'Socket' emerges slowly and with curiosity into the fresh air, a space and a feel that we can assume she has never before encountered, that this signals a gentler pace, heads into another form of decoy for those following the threads of disability, animality and the feminine.

Two gestures in the closing sequence rapidly choreograph the domestic theatre of raising animals. Released from the cellar, the Woman's vertical ascension into the light and into her bloody substitution of her self for the face and the heart of the family contrast with Socket's off-leash ostensible freedom. With two spaces of containment sundered and two captives liberated, together with the visceral shock that one of 'the dogs' was another daughter, one might expect them both to transcend their conditions. However, only the 'animal' in the cellar returns to light, a process already given a toehold with the scalding water: she is lightened as she ascends. But when the Woman encounters Socket, she sees her not as someone to be returned to humanity but as something to be domesticated. This new family is forged – just like the old one – with the sacrifice of animal

Figure 3.3 Pet (*The Woman*, 2013, Dir. Lucky Mckee, USA).

kinship, as Oliver notes.[93] Economically, with one swift disciplinary strike to Socket's head – issued with the hand holding the heart rather than the one with the improvised sword – the woman and *The Woman* retain Socket within an immanent pethood. The sign 'pet' contains the abject threat of Socket, a sign accepted with the reward of the heart, albeit a heart that is now downgraded to the status of 'scraps'. As cast-off scraps, the heart no longer carries the charge of totemic substitution but organizes domestication as a lower order.[94]

The second gesture, also achieved economically in an otherwise speechless scene, occurs when the Woman approaches the house, with Socket now ambling along at her heel. Darlin' rushes out to greet the Woman, to Peggy's palpable terror (and we should note that Darlin' has been consistently addressed with this diminutive 'pet' name throughout the film – never as 'Darleen'). But Darlin' fearlessly offers this *guest* a drink of water in perfect hospitality: she has never recognized the Woman as a hostage. She gulps it down but pays no attention to the blood smeared all over her face and hand, offering her bloody finger to Darlin'. The very girl who had mimed Cleek's injury in the cellar as both victim and perpetrator now licks the blood and smacks her lips. Where unruly bloody issue might more habitually call up the threatening pollution of abject menstrual blood and the stench of that image cannot be erased given the mother's demise, this blood, we know, is that of Darlin's father. This blood is that which has been made to flow; it is the blood of cruelty.[95]

Consanguineous, the Woman, Darlin' and Peggy head off into the woods, their new family replete with pet. As such we must raise doubts as to this feeding of the limits of the Family. Oliver's caution that the potential *trophe* (nourishment) that Derrida affirms within limitrophy can always lodge instead in a trophy such that 'nourishment is always at some level also conquest' stands here.[96] Even if the architecture of the House is sacrificed along with its paternal and maternal figures, the question of the Animal recurs.

Coda: Non-powers of horror

As if in oblique acknowledgement of the murderous character of the film's denouement, the bloody nature of the new Family, and the limited degree to which the structure of that Family shakes the structures of violence that it ordains, *The Woman* has a secret ending.[97] After all the credits have rolled, and thus in a location ambiguously inside and outside the film, an animated cartoon-like fairy tale sequence returns us to Darlin', as if the mobile psychic journey of the polymorphous little girl was or should have been central all along. On a quest, in a little boat at sea, Darlin' sees signs of life on an island that draws her curiosity. As she does within the body of the film, Darlin' greets those that others fear, fearlessly. In this post-scriptum she can give another category-defying entity – a tree-woman-beast living being – flowers, without having to lay them outside a locked door or having her gift forbidden by another. In this case no blood flows, but the two creatures greet each other with a smile.

4

Speculations: Gesture in *Conceiving Ada* and *Absent Presence*

What follows now is Speculation.

– Jacques Derrida[1]

If woman had desires other than 'penis-envy,' this would call into question the unity, the uniqueness, the simplicity of the mirror charged with sending man's image back to him-albeit inverted. Call into question its flatness. The specularization, and speculation, of the purpose of (his) desire could no longer be two-dimensional.

– Luce Irigaray[2]

Give me power *with* pain *a million times over, rather than ease with even talent.*

– Ada, letter to Lady Byron, 25 July 1843[3]

This chapter tells two tales of *telos*. It does so through two differently articulated cinematic fantasies concerning genetics in light of the overlapping writing of Jacques Derrida, Sigmund Freud and Luce Irigaray. Freud's account of the first steps towards language through his grandson's game with a cotton reel and its vocal accompaniment (understood to approximate '*fort*' (gone) and '*da*' (there)) has been a mainstay of psychoanalytic theory and feminist appropriation of that theory.[4] *Speculations* sets aside the extensive literature on *fort/da* that tends to accept Freud at his word, a word that describes the carno-phallogocentric overcoming of human weakness in the world with a string of substitutions to recharge this privation with eventual ability. To wit and in short: the suckling infant loses the constant supply of milk (metonymy of the breast), loses the breast (metonymy of the mother), loses the mother with whom that infant felt in paradisiac union. However, that same infant spurred by these very losses is driven to substitute a toy – the cotton reel – for the absent mother, then to supplant the word for the thing and thus to enter into the hallucinatory pleasures of representation.[5] Instead, the drive of this chapter stems from Derrida's 'To Speculate – on "Freud"', a text that takes Freud at his word in a less literal and more literary sense: rather than accept what Freud says about this game, Derrida is fascinated with what his text does. He is especially fascinated with the gait of the text, so to speak, with the ways that Freud tries and fails to step beyond his ostensible pleasure principle. And rather than simply extracting the brief *fort/da* episode from the maddening non-progress of 'Beyond the Pleasure Principle', Derrida tracks the rhythm of gone/there, *fort/da*, across the whole text as grandson and grandfather effectively play the same game. We can call this game 'autobiography', as Derrida does, albeit that it is played along the altered lines of what he names an 'athetic' writing.[6] As athetic, Freud's texts confound logics of position and opposition, thesis and antithesis. This seemingly familiar and respectable genre – autobiography (the pole position of the self) – will turn out to have more in common with 'science fiction' than we may have expected. Philosopher and analyst Luce Irigaray steps in with yet another rhythm, one that counters both the Freud that she names and the Derrida that she does not, through what she holds to be the interests of the sexually different girl.[7] What is extraordinary about Irigaray's short lecture 'Gesture in Psychoanalysis,' is that she seems to want to both bypass the phallogocentric gambit of *fort/da* through the girl's properly feminine choreography, and also to do what Freud does, in Derrida's reading, namely project a legacy. The projection of legacy as teleological ambition is precisely what is at stake in the two films under discussion here, both of which trade on the vanguardist, sexually ambiguous persona of actress Tilda Swinton.[8] This ambition, however, becomes subject to – *cannot not* become subject to – what Derrida names the 'Postal Principle'.[9]

The first is an experimental feature made by an American artist with a long-term investment in feminism and with a long history of investigating the performativity of female identity through performance and 2D media.[10] This film readily circulates with feminist contexts: Lynn Hershman Leeson's *Conceiving Ada* ostensibly functions as a feminist fantasy concerning the attempt of Emmy (Francesca Faridany), a contemporary American computer scientist, to save a 'forgotten woman' from a history that restricted her genius.[11] Now given an annual 'day' in October, through the frame of a biopic with the added license of science fiction, Conceiving Ada proposes that Emmy has devised a way to speak through her computer interface directly to Ada, Countess of Lovelace (Swinton) in Victorian England.[12] This is not skype. Conceiving Ada is not satisfied with period drama. As science fiction – or 'speculative fiction' – the film crosses time as well as space to produce this 'one to one' with the woman credited with inventing the first computer language, Ada, who died of uterine cancer, aged thirty-six, in London in 1852. Emmy's efforts don't stop at first contact with the past; she banks on archiving Ada to both her computer's hard drive and her unborn daughter's DNA. In spite of Ada's refusal, Conceiving Ada chooses to make the dead live again, there again, *da* again. Given time, she will be able to complete her work; she will be recognized. Ada will be saved. Feminists have long negotiated which strategies to inhabit in order to address the question of women's contributions to various disciplines when historical records falter or finesse their presence. In broad strokes: should we restore female geniuses to their rightful historical places, or should the conceptual grounds of such canonising gestures themselves be deconstructed? My concern here is that Conceiving Ada lauds the former without doubling back to the latter. Moreover the recognition of Ada becomes bound up with a new fantasy of legacy, with a new proper line. Conceiving Ada attempts to seal *tele* as *telos* – the proper ending that was due to Ada. She should have been a progenitor: this will have founded her history. In so doing, will that future have been feminist?

The second is a short film (also exhibited as a five-screen installation), made by a British-based Turkish-Cypriot artist, more prominent as a fashion designer whose clothes often express technical ingenuity (a table metamorphoses into a skirt; a dress can be folded up and addressed as an airmail letter). Hussein Chalayan's science-fiction short *Absent Presence* stars Swinton in an all-female political environment, performing a scientific experiment to demonstrate the ability of DNA to chart origins (thus speaking obliquely to Ada's insights into the future applications of computing).[13] Situating power asymmetrically between women of different ethnicities and nationalities, the film points to our ever-more technicized terror of the other and monitoring of their movements (a fear being weaponized to new extremes at the time of writing). Yet classical science's teleological expectation founders as *tele* takes (its) place: the experiment fails to reformulate the DNA samples of three foreign volunteers as their seamless equivalents. Instead it produces futurist sculptures in which the volunteers struggle to recognize themselves. We never hear those foreign women 'speak for themselves'. This is not an oversight or wanton effacement; rather, the film unravels the ostensibly autonomous white Western subject deploying key tropes of auto-affection. 'Speaking for oneself', subject to the postal principle, is in question. Auto-affection is always and already hetero-affective. Thus, as the genetic 'tracks' of her subjects take unexpected directions, so too does Swinton's authority.

Postal principle

Thought to rest on mastery through substitution, *fort/da* subjects lost objects to being recalled. While 'To Speculate on "Freud"' was published in *The Post Card: From Socrates to Freud and Beyond* in 1987 (translated from the French edition of 1980), Derrida's speculations derive from his mid-1970s' seminars on *Life Death* (*La vie la mort*), in which he directly engaged the theorization of programmes of instruction given in pedagogical or institutional life alongside those given biologically, specifically in François Jacob's work on heredity.[14] 'Life death', as Dawne McCance remarks, 'substitutes for either an *and* or an *is* a silent, invisible "*trait blanc*" between the words "life" and "death"'.[15] This relation will also emerge from *fort/da*, written with varying punctual instructions in 'To Speculate on "Freud"'. This invisible 'treatment' does not suggest that the two terms are 'not two, or that one is the other, but rather that the difference at stake between the two is not of a positional (dialectical or nondialectical order)'.[16] Yet these speculations speak rather more to Freud's projection of his own institutional inheritance and the complicated scene of writing in which he is enmeshed than to genetics *per se* even as that scene itself demands the thinking of 'life death' all the way through. In fact Freud raises the term 'speculation' in order to allow himself 'free rein' and despatch any hint of philosophy, which is of 'no concern' and certainly not – Derrida's suspicion! – the philosophy of speculative dialectics.[17] If it was of concern, and if it did have something to say about pleasure, Freud would 'readily express

[his] gratitude'.[18] As things stand, Freud feels free to speculate upon a cleared ground, without influence and thus without debt. For Derrida, Freud's auto-acquittal serves the purpose of allowing for the inauguration of the entirely autonomous, free-standing, House of 'Freud': psychoanalysis will properly begin with the name of Sigmund Freud. They are gathered together 'under the same roof'. Freud in fact tells us that he has

> been able, through a chance opportunity which presented itself, to throw some light upon the first game played by a little boy of one and a half and invented by himself. It was more than a fleeting observation, for I lived under the same roof as the child and his parents for some weeks, and it was some time before I discovered the meaning of the puzzling activity which he constantly repeated.[19]

Psychoanalysis will improperly begin with 'SF', to use the common abbreviation of 'science fiction' or 'speculative fiction'.[20] In light of Donna Haraway's ironic pressure upon the signature of 'SF', we might also insert 'Speculative Feminism' to this roster ('So Far', as she says, in an indication of a trajectory for which all bets may be off).[21]

Following the way that the form of Freud's writing mimes that of which he writes, Derrida notes the *fort/da* of disappearance/reappearance that Freud himself plays with the pleasure principle. Derrida shortens this principle to PP. PP can then take in the primary processes (that bind the unbound energy that the PP ostensibly wishes to stabilize since 'unpleasure corresponds to an *increase* in the quantity of excitation and pleasure to a *dimunition*').[22] To the Anglophone ear PP is 'peepee', with its twin senses of a little, perhaps child-sized, penis and the urinary fluids that flow from it in a manner that requires training. PP, too, is *pépé*, the affectionate French term for 'grandfather'. Finally and most fatally, the PP cannot avoid the eruption of the postal principle. A postal principle haunts every address: every biography, every autobiography. It is necessary to underscore the auto-bio-graphic because 'Beyond the Pleasure Principle' is a text that, if it goes beyond anything, it goes beyond theoretical reflection. It is not a particular species of autobiography, since Freud drew on that which was close to hand, his nephew Ernst for example, even as he veiled this close connection, as my epigraph demonstrates. In his family house, Freud, '[t]he speculator was not in a situation to observe' with any degree of neutrality (or indeed modesty, as it is discussed in Chapter 6 of this volume).[23] But without quite realizing it, 'Beyond the Pleasure Principle' text writes the conditions of autobiography itself. Ernst throws the cotton reel out of his 'curtained cot': Ernst reels it back again. Repeat. In banking on this return, in obtaining the 'greatest pleasure' in getting a good return, this 'second act' confirms the rewarding logic of re-presentation (*fort/da* in a nutshell).[24] The parallelo-grammatology between the game of the grandson and the writing of his interested witness, grandfather and analyst, does not stop at the particular but informs a general gesture of return: a turn back to the self, the auto-biographical seal of one's own envelope: *fort/da*. However, not only does the principle that Derrida suggests is already at work, require the necessity that a letter (including the self-addressed letter) can always not arrive and thus stay *fort*, dead and gone, 'intentions' notwithstanding, but Derrida's repeated excursions into the matter of origins and ends depart from the anchorage that these bookends are classically taken to secure. Thus not only is re-presentation in question but the notion of an original presence itself. Across *The Post Card*, the post, the letter, address and destination affirm the play of a principle that can never guard against going astray:

> In a word … *as soon as there is*, there is *différance* (and this does not await language, especially human language, and the language of Being, only the mark and the divisible trait), and there is postal maneuvering, relays, delays, destination, telecommunicating network, the possibility, and therefore the fatal necessity, of going astray, etc.[25]

Reduced to its formal structure, the *fort/da* game seems automated, self-fulfilling, easy to cite, to cut and paste: 'The interpretation of the game was obvious. It was related to the child's great cultural achievement – the instinctual renunciation.'[26] There seems to be no question that it should provide the paradigm of language acquisition. It is Derrida who remarks on Freud's assertions of the boy's normality, his good behaviour, good relationships with everyone, that he 'obeys orders not to touch certain things' and 'above all [that] he never cried when his mother left him'.[27] This mother is Sophie, Freud's daughter. No reason to be forlorn. But Derrida is the one to ask, 'Why didn't he cry?'[28] (We can leave the pronoun to its ambiguity.) Already, Derrida suggests an economy balances out both the boy's actions and the grandfather's accounts: the 'child too is speculating'.²⁹ Ernst had a one-track mind, Freud affirms: 'It never occurred to him to pull [the cotton reel] along the floor behind him, for instance, and play at its being a carriage.'³⁰ It does occur to Derrida to ask why: 'why doesn't he play train or carriage? Wouldn't that be more normal?'³¹ He continues to wonder what the PP would prefer. Would Freud have played train? 'What is it to play train, for the (grand)father? To speculate: it would never be to throw the thing.'³² Derrida reminds us that this normal little boy who does not cry does not actually throw away his toy but is able to recall it at any time. The reel is on a string and is thus a safe bet: 'it does not really leave.'³³ '*Fort:da*'.³⁴

The sheer difficulty in maintaining the pleasure principle as such arrives with what is sometimes mistaken for its opposite – the reality principle (RP), while the latter is nothing but a deputy or even, as Derrida notes, a 'domestic' for the pleasure principle put to work when needs must, and needs must when the devil drives.³⁵ The RP ostensibly stands for later, for the lag of deferral, the deferral but not dismissal of the PP, which does not really leave. The question of the return, of making a return, ingrains itself not in an object as such but in 'departure-returning itself': 'the greatest pleasure', 'the complete game'.³⁶ As Derrida notes, Freud's footnote shows the supplementary spool of the child's own reflection as he plays with his own disappearance, one that is never really gone when the mirror will always provide his return.³⁷ The mirror then provides the 'self-addressed envelope', sealed for Freud, forever at risk for Derrida.³⁸ And with this suite of speculative reflections under way, if we 'admit that Freud is writing', he 'does not do *fort/da* indefatigably, with the object that the PP is. He does it with himself, he recalls himself'.³⁹ Something else steps in. The PP is not alone. Freud's 'speculative writing also recalls itself, something else and itself'.⁴⁰ This addition in that which is all additions, additive and addictive, is that which renders this text 'autobiographical'. In so doing, it forges what Derrida calls a 'strange contract' and demands that we reconsider what we thought was the 'entire *topos* of the *autos*'.⁴¹

As PP and RP lose their apparent polarity (reality check: not now, but later), they open onto a difference of a greater magnitude than that classically assumed to lodge in opposition. The back and forth of the text does not simply repeat, put lost objects back in their place. Freud even backtracks on the completion of the *fort/da* game: perhaps the distancing of 'gone' could be 'staged as a game in itself'.⁴² Perhaps it renders a passive situation active. Perhaps there is pleasure in making disappear: a year later this 'same boy' was to be seen by Freud, the disinterested observer, throwing away his toys. He told them, like his father, Freud's son-in-law to 'Go to the fwont!'⁴³ A second footnote – added later and so discontinuous with the writing of the main text – tells us that 'his mother' died when the boy was five and three quarters and that he did not cry then either. The follow-up sentence tells us that in the intervening years a sibling had been born, 'rousing' him to 'violent jealousy'.⁴⁴ Freud states no conclusions about this reaction to the boy's mother's death – that is, his own daughter's death – but the implication is that Ernst's murderous resentful thoughts did not wish for her traitorous return if, in so doing, she would be 'shared'.

This is where, in spite of himself, Freud enacts the 'athetic writing' that so intrigues Derrida. Having the lessening of excitation in common, the PP cannot expel the 'absolute other', that is, the death drive.⁴⁵ There is no fortification against the other that is already 'there'. A demon is already at home: 'the very procedure of the text itself is diabolical. It mimes walking, does not cease walking without advancing, regularly sketching out one step more without gaining an inch of ground.'⁴⁶ This 'not writing', this step of writing, this '*pas d'ecriture*' might frustrate Freud's purported wish to step 'Beyond the Pleasure Principle' but the autobiography that it conveys indicates a general condition for Derrida.⁴⁷ The pleasure principle – like Freud – writes to itself, sends to itself, 'overlapped' by an other.⁴⁸ This overlap gives a new architecture to the uncanny immune to containment by literary fiction.⁴⁹ Overlapped, and not *opposed*, by the death drive, in Derrida's hands the life that ostensibly attaches to the PP is now '*life death*' (the title of Derrida's seminar). It has dropped punctual directive: no forward slash, no hyphen, no colon, no handholding.⁵⁰ Posed in this way, that is to say, without posing or opposing a position, the RP and the PP are *différant*, rather than strictly different, without defence from an infinite deferral. 'There is no thesis of this *differance*. The thesis would be the death sentence *(arret de mort)* of *differance*.'⁵¹ If PP has now become both it and him (quietly in French since both are *il*), 'Perhaps it is that the PP cannot be contradicted'.⁵² This only ironically infers the authority of Freud *qua* infamous patriarch and rather leads to the other track that overlaps the PP, without opposition and without a sound.

Derrida's speculations on 'Freud' bear upon *Conceiving Ada* not only to remark upon the repetitions of *fort/da* that weave its narrative but also because this gesture's investment in returns also ropes in a relation to descendants, to a legacy and to transference. Cutting loose from any philosophical inheritance, these are Freud's own grounds, under one roof. Given that psychoanalysis is being forged as a discipline, its validation requires disciples. It cannot remain private. Home is where the institution is: Freud writes in and of the house of Freud, even as he entertains such a devilish guest as the death drive. From singular to plural, Freud's address shifts between 'I' and 'we'; 'we' projects the legatees who will return all credit to us. The grandson's game – understood as a writing, or just *as* writing – confirms psychoanalysis even as it instructs the grandfather what to write. It is affiliation: Derrida remarks on the confluence of *fils* as both son and thread, strung together. The PP reapplies (to) a line of descent, which would run into trouble if this line was ever pissed away, *fort*, disappeared, died out. '[N]o legacy without transference', says Derrida.[53]

Archive fever

The vexations of *Conceiving Ada* lodge not simply in the over-reliance on the charismatic pull of lead actress Swinton, nor with the evacuation of the epistemological question of how we know the past. Rather they lie in the *fort-da* game enacted by Emmy in retrieving Ada – making Ada *da*; in the transference that the film solicits but never questions between Emmy and Ada as well as between audience and Ada; in the plot that manages the intensity of the relation between two adult women through the heteronormative obligation that one of them – Ada – become the daughter of the other; and for this daughter's conception to require a scene that goes beyond the ostensible digital dexterity of Emmy to include the input of her male partner, Nick. Expressly forbidden to touch Emmy's home computer, Nick, the non-computing expert, goes ahead and does so anyway – to what effect the narrative does not state, but it is hard to read as anything but an insemination (and as reassurance of heterosexuality).[54] These are questions that critical writing on Hershman typically glosses over in lieu of praise of her own technical innovation – the film patented the use of 'blue screen': its period *mise en scene* were digitally composited. Critics draw attention to the fabrication of this film even as the film itself employs well-worn Hollywood practices for masking that very fabrication in the familiar interests of audience identification. Sharon Lin Tay authorizes *Conceiving Ada* as feminist because she understands the film to contrast 'women's loyalty and generosity to one another' with the 'continuous narrative of great men or the perpetual drama of the family romance' and at the same time breaking with the structure and aims of classic realist linear narrative.[55] Tay also determines the film according to a Deleuzian becoming that, she holds, licenses a definitive break from psychoanalysis. Without necessarily dispelling Deleuze, this chapter takes another path. Turning to psychoanalysis does not lead unquestionably to servitude to dogma. This chapter demonstrates that the kind of psychoanalysis that is at stake in *Conceiving Ada* requires a turn to the postal principle.

Conceiving Ada performatively stakes a claim for the contemporary existence – and success of (at least a certain kind of) – liberal feminism: the film opens with Emmy having sex with Nick, her less professionally eminent male lover before turning back to her award-winning work on her Apple (a tiny antique Mac by today's standards, but signalling modernity, creativity, 'coolness' even a hint of radicality in 1997). *Conceiving Ada* too keeps its discipline in the family. Emmy works from home. Home is where the terminal is. Emmy writes computer programming, cutting-edge programming requiring secrecy for the sake of this house. It all takes place 'under the same roof'. At this historical juncture – so recent and yet so very far away now – the 'personal computer' could convincingly provide that privacy: there was no Facebook to archive and marketize her data, no Cloud promising endless amorphous storage, no virtual window for unknown others to access personal information – including the videos of Emmy and Nick having sex from the cameras housed in every room.[56] Emmy too projects a legatee, unleashing her from the past and leasing her to the future, though not just any future. Emmy is an observer keyed to her observations, an interested witness. Through her screen, conceived in its most normatively ideological frame as a simple window, Emmy watches Ada's attempts to write, but without being able to sign in her own name. Ada Byron King, Countess of Lovelace, daughter of Lord Byron and Annabella Millbank, writes footnotes for other people's texts, long, long footnotes for Luigi Federico Menabrea, for Charles Babbage: footnotes that overstep their space, yet signed only A. A. L and liable to be edited, overwritten, co-signed.⁵⁷ Ada has passionate extra marital affairs. Ada writes the first computer algorithm: Ada speculates upon the creative uses to which computers might aspire beyond calculation, beyond number and beyond quantity (thus, in Freud's aspirant terms, beyond the pleasure principle).⁵⁸ Ada panics that she will run out of time, that her work will not be done in time. Ada throws herself away, a wastage repeated by her gambling at the racetrack that started off so well, so mathematically well met, but ends in debt, debts that she could have made good if she had had more time.⁵⁹ Ada is frequently ill; her pain is held in check with opium but she bleeds too much, and, in an age of 'curative' blood-letting, continues to bleed all the way to death. Save, Emmy puts a stop to this overspill and transfers Ada to a place in which archives are without fever, a fortified future perfect. The computer-programming language devised by the US military and named after Ada is not the repository that Emmy has in mind. Save, Emmy restores a lost 'mother' of mathematics: but Ada is found, or founded, as her daughter: kept in line, no more a legator but a legatee.

Irigaray's legacy

As presented by Irigaray, Freud's story of Ernst's steps towards language and 'the symbolic order' is not a transferable, universally translatable account.⁶⁰ It is an account of the boy's experience, generalizable to that of all boys. 'Beyond the Pleasure Principle' is thus consistent with Freud's essay on 'Femininity': it repeats the absence of sexual difference such that there is only phallic masculinity and its negation, otherwise referred to by Irigaray as the Same and identified as the dialectical habit of the history of canonical Western philosophy in *Speculum of the Other Woman* (the substantial volume that famously led Jacques Lacan to expel her from his school of psychoanalysis, *L'Ecole Freudienne*).⁶¹ For Irigaray, the girl remains unaddressed (in the double sense that she is not addressed or spoken to, and she has no address of her own). If Irigaray's own 'Speculations' on 'Freud' in the vast opening chapter of *Speculum* named 'The Blind Spot in an Old Dream of Symmetry' have been of silent benefit to Derrida (as Amy Hollywood suggests) that silence is returned in her 'Gesture' (i.e. neither names the other even as the sexual poetics of their work often overlap).⁶² One could surmise that she has simply not read the essays collected in *The Postcard*, yet her earlier essay called 'Belief Itself', first delivered as a paper at *The Ends of Man* (the 1980 Cerisy-la-Salle colloquium in honour of Derrida), amply demonstrates this not to be the case.[63] In the 'Gesture' essay, when Irigaray mentions raising 'this question of Ernst's maleness one day at a Cerisy conference … someone objected that Ernst could have been a girl'.[64] Irigaray's answer was: 'he was a boy … it couldn't have been a girl. Why? A girl does not do the same things when her mother goes away'.[65] In the 'Blind Spot' she had previously remarked, '*No fiction, no mimetic game, is allowed the little girl if it involves herself or her relationship to (re)production.* Such games are "phallic"'.[66] In 'Belief Itself', Irigaray opens with the pointed concern that the girl, that sexual difference in the feminine, is detained at the '"poste restante" or P.O. Box where messages for unknown persons with no fixed address are held, undeliverable by the usual, already coded, telecommanded, circuits.'[67]

It might be thought remarkable that, for *The Ends of Man*, Irigaray chose to address Derrida by means of their close mutual reading of the *fort/da* episode rather than his deeply critical reading of Lacan in 'Le Facteur de la Verité'.[68] In that long essay, arguing against Lacan's formulation of signification in his essay on Edgar Allen Poe's *Purloined Letter*, Derrida infamously wrote: 'Not that the letter never arrives at its destination, but it belongs to the structure of the letter to be capable, always, of *not arriving*.'[69] Far from a merely formal disagreement, Derrida's challenge to Lacan, whom he names the 'postman' or 'purveyor of truth', was that in reading the itinerary of this stolen letter, he predictably finds it 'between the legs of the fireplace' and lodges it within a classical suite of femininity, truth and castration.[70] Derrida continues, by way of contrast: 'dissemination threatens the law of the signifier and of castration as the contract of truth. It broaches, breaches [*entame*] the unity of the signifier, that is, of the phallus.'[71]

Irigaray's desire for her 'letter' to arrive cannot be confused with that of Lacan, since the latter's arrival would be in the form of what she calls the Same (the negative image of the phallus rather than any kind of positive difference). That Irigaray's letter has a red 'signature' again refuses this same appropriation: 'Belief Itself' begins with a brief account of a dream from one of her female analysands, a story that this analysand had effectively withdrawn from another analyst on grounds that no useful interpretation emerged (it was stuck at the *poste restante*). In the dream, when the father and the son enunciate the words of Holy Communion – 'This is my body, this is my blood' – the woman bleeds. Rather than repeat the scene – the seen! – of the bloody wound of castration, this dream might be consonant with Irigaray's own frequently cited desire to jam the machinery of converting the sensible into the patriarchal economy of signs.[72] Rather she will come to insist on a 'sensible transcendental' in which flesh and word come together without losing touch with a feminine sexual difference that cannot be said to be 'binary' or 'oppositional' in any sense. What might appear to be a frontal criticism of Derrida – the implication that he too runs the risk of repeating the negation of the feminine even as he plays with sexual differences – becomes a complex step in her dance, her descendance, even as the feminine sexual difference of which her work has often spoken is *not one*.

When Irigaray takes Freud at his word in her 'Gesture in Psychoanalysis', the substitution of reel for word for mother for self is emblematic of the boy's compensatory manoeuvre when encountering her sexual difference. She does not refer to the context in which Freud describes this game other than the bare bones of Ernst's use of the reel and string when his mother is away. His 'house' is not in view in her account. Doubling Freud, she draws on her own clinical experience to sketch the girl's movements, speculating that these are not the same as those of the boy. Stating that she gives no specific case history account in order to 'respect the ethics of psychoanalysis', Irigaray says that she will refer only to 'certain movements and actions that occur in every analysis'.[73] These movements refer principally to the classical architecture of the analytic environment that privileges a dissymmetrical relation of patient to analyst – one that is not face-to-face and is brokered by the couch on which the analysand reclines. 'Disobey[ing] not only social convention but also the relation of signs to language', the scenography breaks with attention to the present and solicits reminiscences.[74] The scene is set to work upon transference, to push for the past to erupt in such a way that it cannot be mistaken for the present – and the relationship with the analyst who has nevertheless generated this repetition in order that it might be worked through. Irigaray also notes that the asymmetry between patient and analyst, one lying and one sitting down, is a gendered formation liable to register hierarchically organized sexual connotations for the men and women that inhabit them.[75] Given the dogmatism of the horizontal as the plane of the animal and the vertical that of man apparent throughout this book, we might also understand this scene as a species of domestication (emblematized by the order to 'lie down').

Unlike her own decorum in never revealing particulars and persons of her analytic practice within her philosophy publications, Irigaray writes with some ambivalence, 'Some of my male and female patients have written for themselves, without even mentioning the name of their analyst in most cases. It is true that I am a woman. Which must partly explain why I am not cited.'[76] She is not recognized, has not been addressed: she has been barred from descendance, left in the *poste restante*. Moreover in the rare circumstance when she has mentioned 'a few fragments' detailing her practice, she doubts whether any of her patients could 'recognize themselves' anyway.[77] Again her tone is ambivalent as to whether we can assume that Irigaray congratulates herself on her ethical discretion on refusing to 'eat the other' for the sake of her own subsequent writerly exercise (and that congratulation could vehicle a dose of the PP).[78] Following her description of the gestures of every analysis, Irigaray gives further attention to the case of Ernst/the boy in general, and then turns to the girl. We are never told from whence this girl derives. But 'the girl' proliferates into 'girls' as the essay steps along. She, Irigaray suggests, is not confronted with an other in the mother, and thus there is no other to be managed by introjection or incorporation. In the next sentence girls bear witness to the girl in mitigation against mere speculation: 'on the contrary, girls often set up a defensive territory that can then become creative, especially in analysis.'[79] We can speculate that these girls reflect the totality of her own clinical experience and thus reflect Irigaray's view of girls in general as her discussion moves into recognizable figures from her previous writing (such as the lips as one possible metonymy of the sex which is not one).[80] She further draws us in as witnesses however, to the evidence that girls display in the 'school yard' in which they

> describe a circular territory around themselves, around their bodies … Sometimes girls whirl around in silence or else they giggle, and chatter, and chant nursery rhymes. Perhaps chant is not quite the right word: they make up variants, invent phonic and syllabic games.[81]

Regarding Ernst, Irigaray elaborates an aspect of the game that neither Freud nor Derrida pursues: the oral gestures made as these '*o-o-o-o*' and '*da*' sounds are pronounced. In her account, the '*o-o-o-o*', understood by Freud to foreshadow '*fort*', escapes the mouth without being held back or swallowed but comes to a halt with the 't'.[82] *Da*, the dental consonant, in contrast, is swallowed. The sound of closeness is introjected; the sound of distance is limited and hence mastered. Whether or not this limited incorporation holds – and Irigaray herself raises the comparative question of languages other than German – Irigaray fleshes out the boy's activities indicating the use of his arm. Ernst both vocally and physically gestures towards his absent mother (without walking or crawling towards her, as Derrida also remarks). It is as if Ernst is 'listening to himself' in this early linguistic manifestation, says Irigaray.[83] Again she echoes Derrida but apparently without his destinerrant loophole. Speaking, gesturing, returning to himself, it is also 'as though Ernst were driving a car or a tank', driving 'with his mouth, his string and his reel'.[84] Banking on a journey that will always return home, she notes that Ernst also plays the game with his mirror image: his image may disappear, but he can always reel it in. Or he believes this to be the case. For Irigaray, *fort-da* is the vehicle of that belief 'that the disappearance-reappearance, inside-outside, outside-inside can be mastered, whereas in fact they can no more be mastered than the life-death watch that is our obligation from birth, if not before'.[85]

The girl, we learn, deals with her 'anorexic' distress at the mother's absence without substitution (and hence mastery). On the one hand, she dances and in so doing creates a space of separation, a territory, for herself. On the other hand, she plays with a doll, again in the interests of developing her own symbolic space, and again in a fashion distinct from that of the boy. Irigaray firmly draws a distinction between dolls and boys' toys. Dolls may assist with the formation of a symbolic space but they themselves do not symbolize. The symbolized object has become just that – an object, and Irigaray has it that 'the mother is a subject that cannot easily be reduced to an object' for the girl.[86] This is apparently because she already knows her sexual commonality with the mother in contrast to the boy's formation of sexual difference that leads him to objectify and instrumentalize her. Significantly, that commonality is based on the auto-affective 'not one' that blurs the relation between inside and out. While the girl may sing melodically to the doll or alternatively address angry words to it, Irigaray never compares this to Ernst's subsequent vocalizations: the girl never banishes the doll with a 'Go to the fwont!' For the boy *fort-da* is defensive:

making mother present again 'protects him from disappearing into her. This return annihilates that other return that might swallow him up, take him back into that first dwelling place inside her'.[87] In phrasing it thus in 'Belief Itself', she places the boy in the frame of speculative dialectics (annihilating the other, leaving no remainder). For the Irigaray of 'Gesture', feminine sexual sameness mars the mastery of the game of *fort-da* and thus grows out of but not away from the sensible.[88] As the same, the two of them (mother and girl) can never be decisively parted. When the girls dance, they express a spiralling or 'whirling' sexual gesture in distinction to the *fort-da* that Irigaray identifies as both linear and the analogy of both sexual penetration and its masturbatory substitute.[89] Irigaray identifies this 'whirling' as the 'sexual movement characteristic of the female'.[90] Later she says that they 'whirl not only toward or around an external sun [an external point] but also around themselves and within themselves'.[91]

In spite of the critique of representation at work in Irigaray in common with Derrida, is there thus, nevertheless, a repetition of *fort/da* in her formulation? All the girls demonstrate the feminine of which Irigaray would write. All the girls confirm the House of Irigaray, just as Ernst confirms that of Freud, even as they are both in her analytic experience and in the school yard of our own everyday lives. They are her descendance. They dance (with) Irigaray's legacy (remembering that Derrida puns across languages on the 'Legs de Freud' at some length, with the leg and the legacy further embodying the problem of how Freud's text walks and/or works [*marcher*]).[92] Irigaray emancipates their movement: feminine subjects need 'to be free to *walk*, walk away and walk back, however it pleases them'.[93] The 'paralysis' of hysteria 'strikes' her legs when her proper autoeroticism is inhibited (when the girl cannot whirl).[94] But the girls' steps should not suffer the drag of Freud's inability to take one step beyond the pleasure principle. Likewise a future philosophy of the feminine should suffer no erasure.

Citing Derrida in 'Belief Itself', Irigaray includes several paragraphs of his interpretation of Freud's account of Ernst, including that he was a 'good boy'.[95] Her citation, however, begins after Derrida has queried whether Ernst was a 'normal' boy and thus could really stand as a 'paradigm' for human language acquisition.[96] While Derrida and Irigaray frequently overlap as we have seen, could we say it is in *Irigaray's interest* to secure the boy's position to better offset the girl, the girls, her girls, the chorus line of her legacy?

Derrida brings Freud's principles into contact with each other to the ruination of position – and thus the discrete field of representation – in and through the labours of the boy, which are also those of the grandfather, producing a general condition other than the one the PP tries to render apparent. From within its own grounds, the paternal line begins to reel. For Irigaray, on the one foot, Ernst remains The Boy, hell bent on mastering the mother and nailing sexual difference as that between the one and the 'nothing to be seen', the exemplar fetishist.[97] On the other foot, the girl generates the contact zone of temporary, temporizing and spatializing moves that improvise distances and fields. In so far as this whirl is the open-ended gesture breaching inside and outside without such openness ossifying into a wound or woundedness, perhaps this could become the general condition of *sexual differences*. A footnote in 'Belief Itself' issues a warning against reinstalling sexual difference as opposition: 'all phallic norms make sexuality devilish in the sense that this order is interposed to blur the sympathy between the sexes.'[98] The devil in this detail is the guardian of opposition. It is not the demon revenant in Derrida (and Derrida's Freud). There, no, *there*, it

> comes back [*revient*] without coming back [*revenir à*] to anyone, it produces simple effects of ventriloquism without origin, without emission, and without addressee. It is only posted, the post in its 'pure' state, a kind of mailman [*facteur*] without destination. Tele – without Telos.[99]

That demon is of no fixed abode. What might bypass attention is the sympathy between those non-oppositional sexes that Irigaray's footnote also lets slip. This sympathy, or affinity, might both disarm those critics who would still cry 'essentialism' and depose Ernst as the masculine envoy.[100]

Migration

Conceiving Ada allows Emmy to send her artificial life forms or 'agents' into the past in order to retrieve information. Affectionately known as 'Bird', Ada is given a bird, a mechanical bird.[101] Locating a notional point of origin of Ada holding this bird in a photograph taken by Babbage, *Conceiving Ada* speculates that Emmy can save, hence copy, and then deploy this bird much

Figure 4.1 Bird (*Conceiving Ada*, 1997, Dir. Lynn Hershman-Leeson, USA)

like a homing pigeon. Let loose, a homing pigeon will only return to its source, its 'home'. But as Emmy's copy, the bird is a double agent; it has a second home and thus returns to Emmy. The bird – which bird? – can be sent away (*fort*) in order to home in on targeted addresses within Ada's life (as vague – or perhaps as precise! – as 'Ada's origin', cue the primal scene of her conception). A bird for a bird, it is sent away in order to return home (*da*), having retrieved the information as requested, missive accomplished. This pet-name pet-bird has a specific flight pattern. On automatic pilot, she cannot trace a random flight through the past or score her flight with one wing while erasing it with the other as with Freud's gait in his attempt to step beyond. No, the greatest pleasure is in the return(s). Who Emmy? In the instant that she interrupts Ada's present as an invisible voice from the future Ada abruptly asks, as if to us, to the audience in the future, 'My mother always told me I had a devil or an angel watching me; which are you?' Emmy says that she is 'a friend'. Emmy: *Amie*. Or Emmy: *Ennemie*.[102] It's a close call. Friend enemy, angel devil, mother daughter. In Irigaray, the devil

> blurs the future and any hint we may receive of it ... Only doubles are present and represented, only reproductions, kinds of negatives, reduced prints-only the angel can give light and expansion ... Everything seems to be programmed, predictable.[103]

What is predictable is a safe bet. Ada repeatedly asks, 'can you save me?' Her plea ostensibly replaces a religious figure of redemption – to be saved from her loose ways (loose with money – with figures, debts, speculation) and to be saved from her uterine cancer (blood loss on all sides), with digital archivation (being saved in digital storage and the promise of everlasting 'life'). Emmy responds as her fingers grace the 'return' key: 'yes, I think I can.' In the film's closing sequences the gilded bird cage that Nick has given Emmy is in shot: inside it sit two stuffed birds.

The Cartesian automaticity of the bird's return repeats – even when, *diabolically* when, on her deathbed in the penultimate scene, Ada finally says, no, 'do not save me' and that she does not want all her secrets known, does not want to be a searchable database.[104] Going forth at the end of the film that she expressly wishes to be her end, Ada 'wishes to die only in [her] own fashion', to paraphrase Freud.[105] Derrida repeats Freud in stronger terms: regarding the death drive, the organism wishes to 'die properly, I am living so that I may die properly, and so that my death is my own'.[106] Elsewhere in this book the pernicious end game of the animal question is explored to the effect that within this tyrannical conceptual field, the animal cannot die (a thesis most famously brought to attention by Derrida in several critical accounts of the work of Martin Heidegger). In her work into the archives of Western thought

Figure 4.2 Devil Angel (*Conceiving Ada*, 1997, Dir. Lynn Hershman-Leeson, USA)

regarding death, including its most canonical formations in Hegel, McCance finds that women too have been philosophically excised from a 'proper' death and are held to merely 'wither away'.[107] Thus, McCance is attuned to the replay of the improper death of women in Derrida's *Life Death* seminars, identified in the genetics of Jacob. In that context, Jacob imports a feminized relation between bacterial cells in order to argue that death is proper only to the internal capacity of the cell, and in the absence of such capacity, the cell does not properly die. Consequently, 'Jacob says of the mother bacterium that, once her DNA passes to her daughters, she disappears but does not die.'[108]

But Emmy and her agents can even tap Ada's unconscious. Unseen like a Freudian analyst, Emmy speaks like one too: 'Tell me what you remember most', says Emmy the analyst, the digital archivist-programmer. Ada remembers lessons with one of her first teachers of mathematics; the above board formality doubled by a lesson of 'footsie' under the table and her child self's voice-over speaking of her confusion between the passion for learning and sexual passion. Prompting Ada's transferential recollection, Emmy's own counter-transference remains unchecked. And since we are given Emmy's viewpoint, our own transference is also solicited – she really is speaking to us, appealing to us – not least when Ada breaks the fourth wall of the film (in a move that is something of a signature for Swinton).[109] *Conceiving Ada* is frustratingly silent on Emmy's decision to save Ada regardless (as is extant criticism of the film). Oddly, making Ada *da* means making her *fort*, keeping her at a distance. For she is not simply made present, there! – out of thin air, or even downloaded into the computer as a document, but sent 'into' Emmy's child: the child that can *have* access to Ada's memories rather than simply 'be' Ada. Ada dies, goes away, *fort*, but, returns in or as the child, as Emmy's descendant, a guaranteed legatee. Indivisible, Ada has not been lost.

Palindrome[110]

Who Ada, who dada? From PP or *pépé* to dada, these syllabic doubles recall the ethical dilemmas of gender in the West after the emancipation of the vote and the entry of women into public life into which Mignon Nixon has suggested that feminist art might intervene.[111] Even without the patrilineal fealty exacted textually in Derrida's reading of Freud, Nixon notes that the latter's essay 'Schoolboy Psychology' positions the interweaving of culture and history as that which is effected by the transference between fathers and sons as they manage their ambivalent relations (and as evidenced in Chapter 2 of this book).[112] The pre-emancipatory model of father-daughter relations 'stunt[ed] ambivalence rather than cultivating it', she observes, since there was no room for women to rage against their fathers.[113] However, emancipation attendant upon first-wave feminism began to put women in the same authoritative positions as men and thus spurred women into effecting the same kinds of transferential exchanges, thus impacting on history under the same terms. For Nixon, the murderous tenor of the ambivalence inherent to transference is of core concern; this necessarily includes the 'affectionate' cases in which 'hostility shadows the singular esteem in which the teacher or analyst is held'.[114] We might pause on this shadow in light of what seems like a purely affectionate transference between Emmy and Ada. 'Schoolboy Psychology' directly echoes the cannibalistic murder of the primal father by the band of brothers of *Totem & Taboo*, published one year prior. Nixon reminds us of the absolute authority of the subjects that this cultural transference negotiates and asks whether the critical relation to authorial coherence in recent art made by women might inscribe a 'new edition' of transference. It is, however, difficult to see *Conceiving Ada*'s inter-feminine exchanges in a particularly new light, speculative technologies of memory transfer notwithstanding (indeed the film is at risk of performing a 'new edition of an old desire').[115]

The mothers of Ada and Emmy are played by the same actress (Karen Black), and played in the same deeply unpleasant unsympathetic fashion. Worse than that, while Ada is presented as being born in the wrong time, the film saves its specific terrors for the maternal.[116] While Victorian England may have preferred childbearing to mathematics as woman's place, the film casts the mother as the one who destroys her daughter's letter and metonymically her uterus. When Ada's mother intercepts this intimate letter, written to one of her lovers, it provokes Ada's refusal to be saved by Emmy – the one who will be her future mother after all. Who is it that is 'conceiving Ada' anyway? On her deathbed with her mother in shot through the mirror, doubling Emmy's screen, Ada complains that her mother 'knows everything' and that she feels as if her mother is trying to 'suck her back into her womb' no less. 'No wonder', she exclaims, that her 'womb's dissolved'! No more generation: no gesture save annihilation. Both mothers call to her and Ada addresses the 'angel devil' (with what punctual mark we cannot tell). It's the maternal return that takes the rap for death, while the text, misconceived as paternal permanence, proposes a fortress archive. 'Do not save me': delete me. Between the devil and the deep blue sea, Ada chooses death. 'Think of your heirs', Emmy implores, with some ambivalence to this abrupt rupture in their transference. Ada refuses again: 'They will have to take their own chances, like I did, life itself is a gamble, besides I don't want all my secrets known or to be watched after I am dead.' Ada's final words open a loophole: 'Death makes the fragility of life delicious: in general I am not opposed to it.' No, Ada is not opposed to death, *life death*. And/but the hard drive of *Conceiving Ada* already has her, mimed by its digital form.

But the present is a good place to which to return: there are only proper deaths at the appointed hour, are there not? And Emmy is a good mother, is she not? The closing scene flashes forward to domestic bliss between mother-daughter and daughter-mother. It may be avant-garde, a hip artist's warehouse, but *Conceiving Ada* remains homeward bound. Emmy doesn't force hours of mathematics on young Ada: they can differentiate between playing and working. Nick is still there, somewhere. There, there. Emmy lets Ada play with her memories on the computer. But didn't Ada's refusal turn on not wanting everything known? Young or new Ada – what to call her? – has total recall, and Emmy watches too. Again the time period allows the fiction that this digital archive retains a private signature, can remain secret, even as we know that Emmy has already accessed the past and saved it for her personal interest. Earlier in the film when Ada remembers her life at Emmy's analytic prompt, the voice narrating the interweaving of sexual and intellectual passion is shown, startlingly, to have been that of this child. Is this an other, Irigarayan relation, adequate to the feminine? On the surface, it appears to figure the girl's recollection (*da*) of a primal scene without anxiety listened to by a mother-analyst who gives her approval. There are only angels here, *Conceiving Ada* would have us believe. Yet, in spite of this affectionate tone, at the same time, the shadow of cannibalism haunts their relation as the biopic of Ada, Countess of Lovelace, becomes an auto-biopic.

Absence presence

The viewer may not notice at first, but the beginning of *Absent Presence* plays backwards. Even though the film is digital and thus there is no celluloid reel to actually be rewound, the trope is nevertheless at work especially since the mirrored image at stake doubles the process of winding, unwinding, rewinding, giving the sense of repetition, losing the sense of an end. What would securely bind the primary processes? In this film Swinton plays a scientist (referred to in the credits only as the 'Operator') whose work is bracketed by what appears to be a daily ritual of binding her long hair into complex and headache-inducingly tight woven braids prior to work and then unravelling them at the end of the working day. We are never told on whose authority she conducts her work, with initial automaton-like obedience. Playing this sequence of Swinton unbraiding her hair in reverse gives the impression she can perform the unlikely manoeuvre of 'upbraiding' it, so to speak, unaided, autonomously (a fiction that digital film is ideally placed to finesse). This elaborate procedure presumes that no stray filament will infiltrate and thus threaten to tie the

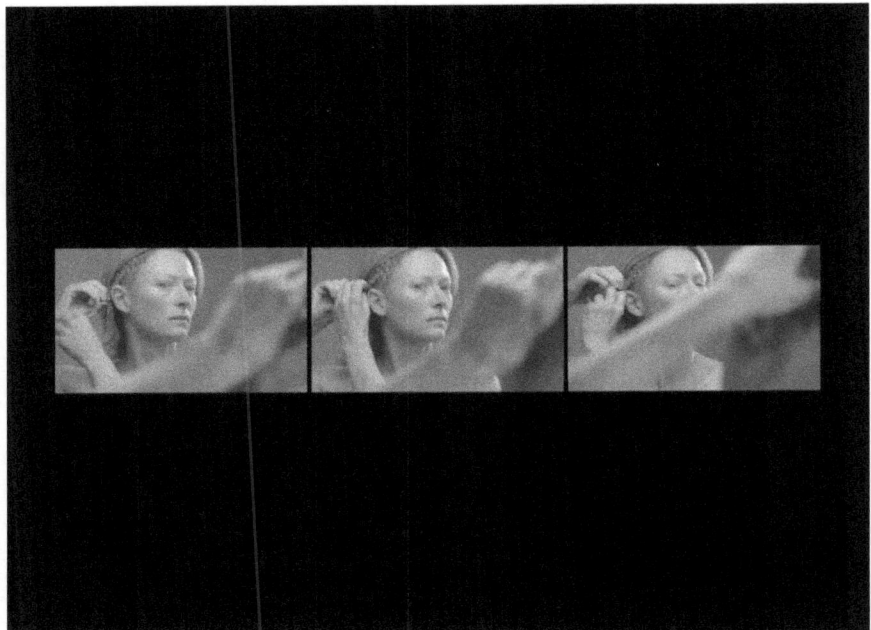

Figure 4.3 Braids (*Absent Presence*, 2005, Dir. Hussein Chalayan, UK/Turkey)

Operator into the operation, diluting its results. Severe, but as Freud tells us of his grandson's game, the greatest pleasure is in the return. As Derrida puts it, 'The movement of reappropriation is the most driven drive.'[117] Ostensibly at work on the others, she must maintain herself (as such). If only the binding could continue *ad infinitum*, and she has enough hair to suggest as much, the Operator would remain unruffled, untroubled by any excess of energy, on the even keel of the PP. She would keep coming back to herself, only writing to herself, keep addressing the postal principle of her own death.[118] But pleasure: it's a bind; the Operator comes undone.

Absent Presence doesn't content itself with an easy mirror image of Swinton's toilette opening and closing the frames of this short film. Crucially the tone of her voice-over, the style of her address, contrasts these sequences, bringing the voice as a key operative of auto-affection into play. Unlike the Operator's hair, which readily suggests its filamentary propensity for entanglement, voice seems to give nothing away. As Derrida informs us in his early writings, voice appears to have no material body – it appears to not appear – and thus to belong with absolute propriety to the speaker: 'My words are 'alive' because they *seem* not to leave me, *seem* not to fall outside of me, outside of my breath, into a visible distance; they do not stop belonging to me, to be at my disposal "without anything accessory".'[119] It is thus liable to support the fantasy of presence. Use of voice-over in cinema traditionally colludes with this (dis) appearance. *Absent Presence* attends to this tendency and, with Derrida, repurposes writing, trace, *différance* in the voice.

The image of the Operator's scrubbed face and bound hair is brutally succeeded by a close-up of soapy water, in which we will learn she is washing clothes. If there is anything odd about the laboratory *mise en scène*, it is the portrait of the Operator on the wall. Not a simple indexical identification of this worker, it interrupts the apparent realism with an address. This portrait shows her with index finger raised to her lips: Be quiet? Don't tell? Mum's the word? To whom is this silent admonition addressed if not the Operator herself?

At first there is only the sound of a ventilation system, giving texture to the air. When the image cuts from Swinton's hair reparations to the austere grey workplace, a diegetic radio is airing what sounds like a play: a man asks, 'more frightened than a sighted person, you mean?' Then the Operator's voice-over calmly strikes up and supersedes the radio. Beginning, 'It was decided to collect clothing from a group of anonymous women donors', the Operator's neutral third-person scientific report relates an experiment designed to trace how the women 'came to be who they are from traces of their DNA left behind in their garments'. The donors are known only as women from outside London, and outside England. Reporting in this form is designed to betray no bias, to erase a merely subjective inflection, adding nothing. It aims for the report to faithfully reflect the experiment, point for point. In this biopolitical environment her crisp enunciation even reflects the eugenic drive to 'link … reproductive control to the production of "proper" phonetic speech'.[120] Indeed, as *outsiders*, the experiment evokes the elementary condition of racism directed at the foreigner whose genetic makeup is under scrutiny. The Operator continues: 'How much could be divined about the women, simply from their genetic makeup?' The theological tone of 'divination' marks the unquestioned confidence in the evidence in preparation, the anticipation of authority. But the change in tone occurring in the final scenes is not simply about reinstating an individual – returning the person to an impersonal voice. Rather the whispering, multitracked, hesitations with which the Operator's voice-over concludes, exposes the divisibility interior to voice itself; its so-called 'living presence' vacillates even in the 'first person'. This secret place of self-address turns on itself, and it mirrors the undoing of all the mirrors – all the returns – in this film.

What sets off this derangement? At the point at which the voice-over starts to explain something specific about genetic origins, geographies and biologies, the Operator's voice multiplies and her voices are not synchronized. Phrases such as 'even more accurately' surfacing in the voice tracks are counter-posed by the loss of any single sentence to audibility. Briefly her single voice-over resumes: 'Now we extrapolate and reduce the chain of DNA which makes her her and subject it to the test of urban living.' But this test is not visualized. Instead the Operator's apparent frustration with the technology surfaces. The SF genre imagines a 'wet' computer interface, which splashes everywhere when she strikes it, rendering a 'dry' inversion of the opposition circulated in early cyberpunk novels in which the 'hardware' of technology was counter-posed by the 'wetware' of the body. This is intercut with what seems like an anxious fantasy in which the Operator, repositioned as a student in a lecture theatre, repeatedly asks: 'how am I supposed to know?'[121] Positioned as the subject of

Figure 4.4 Foreigners (*Absent Presence*, 2005, Dir. Hussein Chalayan, UK/Turkey)

knowledge this worker-scientist may be heard as entering a painful irony of epistemology – *how*, really, are we supposed to know?[122]

Worse: the experiments do not follow the predicted pattern. When the three volunteers return, their three interviews with the Operator are screened simultaneously and the voice-over announcing where each one really comes from is played simultaneously, magnifying the mismatches. Albeit muffled, it is just about audible that the 'individual presumed to be Chinese, was Korean', and similarly that the one supposed to be from France was from Slovenia, and that the urban and rural also cease to align. The three women all speak to camera, but their voices are edited out. Rather than charting a story of origins and of ends, they do not return to sender, and when the clothing is further 'subjected to the sound frequencies of London' and fabulously digitally reconstituted, futurist sculptures stand in place of the outsider, foreigner women. With the appearance of these sculptures comes another style change in the Operator's voice-over. Now she whispers with palpable unease: 'how could we reduce her to this?' We never learn how the women view their erratic results but their distant inspection does not convey the pleasure in returns.

The last sequence has the Operator repeating anxieties about the safety of water as she unbinds her hair: 'Water is a currency of infiltration', her interior voice whispers. As the reflective liquid surface of the Operator's computer, water was not supposed to add anything. As the medium in which the clothing is washed, it should only siphon away, not encipher. Water has ceased to come clean. Looking dishevelled, she splashes water back over her face, again and again. The film dwells on this in a fashion beyond function. It is the most repetitive gesture in the entire (short) film. It is noticeable that *Absent Presence* refrains from posing the encounter between the Operator and the outsiders in the most habitual gendered and racist terms of sexual threat (as a violation, as an intrusion and as an impending imposition upon the fiction of genetic purity perpetrated by male foreigners against 'our' females). Ending without further speech to remark a position or programme of the Operator, any further investment in purity has been dissolved by virtue of the disseminal water. While that investment has operated at the level of the form and assumed neutrality of the Operator, it gives way to a sensual, sensible condition in which we are all 'wetware'. Our vulnerability is staring us in the face. As Derrida wrote on the first page of 'Plato's Pharmacy':

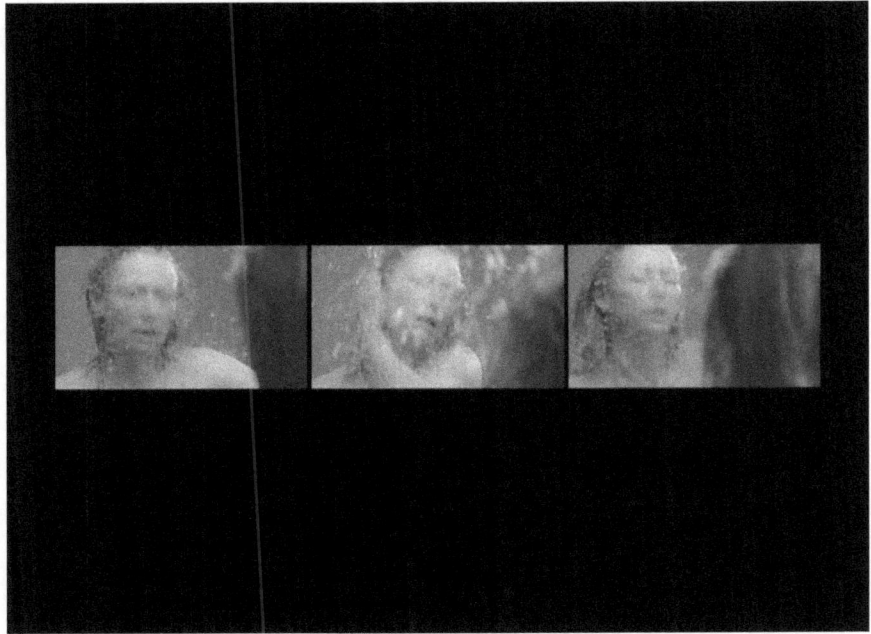

Figure 4.5 Splashback (*Absent Presence*, 2005, Dir. Hussein Chalayan, UK/Turkey)

There is always a surprise in store for the anatomy or physiology of any criticism that might think it had mastered the game, surveyed all the threads at once, deluding itself, too, in wanting to look at the text without touching it, without laying a hand on the 'object', without risking-which is the only chance of entering into the game, by getting a few fingers caught—the addition of some new thread.[123]

The double gesture

In the space of a concluding section – of coming to an end with its implied synthesis – it is impossible to close this chapter, only to speculate with the odds unknown. Who is Ada? *Is* Ada? *Conceiving Ada* holds its tongue concerning the return of the dead, says nothing about this ostensibly feminist new 'publication', new edition of Ada, even as the indefinite article lapping at Ada's name, either side, continues to sign *life death* together. She is a kind of *revenant* about whom it is difficult to decide if she is dead or alive. Are there two Adas? Mathematical Ada, the adder, the one who added up the calculations of others and found them wanting; now Ada added to Ada, data aiding Ada, shored up as if without risk of any subtraction, a good bet. Unbraided at the end that is the beginning, the Operator of *Absent Presence* is master of no one, including herself. We might rename the film *Absence Presence* to take the first term out of adjectival submission: life death. Perhaps the Operator is not one, to second Irigaray. Even as this film set about without any express feminist expectation, the Operator has been touching herself all this time.

In anticipation of the following chapters of this book, we might pause on a speculative thought, as suggested by Dawne McCance when reading the *La Vie La Mort* seminars in light of *The Animal That Therefore I Am*. It is also an affirmative thought as it slips outside of oppositional logic. She writes:

> if the 'death drive' is to be conceived as a 'force' inherent in all organic life, it might be understood as 'vulnerability' or 'passivity' rather than as the 'capacity' or 'power' that ... is said to elevate human over nonhuman life.[124]

For as soon as the ability of instinctual renunciation – the effective, time-sensitive deferral of 'not now, but later' in the activity of recall supposed by *fort/da* – was put into question, Derrida's long deconstruction of the question of the animal troubled the horizon of phallogocentrism.

5

'*Unfamiliar Unconscious*': The performativity of *Infinity Kisses*

What is important to me is not the appearance, it is the passage. I like the word passage. Pas sage (ill-behaved, unwise). All the passwords all the passing and boarderpass words, the words which cross the eyelid on the interior *of their own body, are my magic* animots, *my animal-words.*

– Hélène Cixous[1]

This chapter revisits the figure of the interspecies kiss in works by Donna Haraway, Hélène Cixous and Carolee Schneemann in light of the question of the animal posed by Jacques Derrida. The return named in this revisit is to work I first published in 2010, work arising from the dynamic force of what was then an essay rapidly gaining traction in animal studies: Derrida's 'The Animal That Therefore I Am (More to Follow)'.[2] At that time it had become clear that Derrida had so altered the intellectual frame as to make Schneemann's series of works named *Infinity Kisses* something that could be affirmatively approached rather than avoided out of embarrassment and that the parallel projects of Donna Haraway and Cixous would further this approach.[3] Indeed, the public consolidation of the work of Haraway's 2003 *Companion Species Manifesto* in the extended form of *When Species Meet* in 2008 demanded engagement, and an important issue of the journal *parallax* in 2006 called 'animal being' – itself announcing the extensive development of work in animal studies now engaging continental and not predominantly analytic philosophy – included an intriguing essay by Cixous announcing 'The Cat's Arrival'.[4] Extending my

---

'Unfamiliar Unconscious' is Donna Haraway's term. See Donna J. Haraway, *Modest_Witness@Second_Millenium.FemaleMan©_Meets_ Oncomouse*̈ (New York and London: Routledge, 1997), 265.

arguments from a decade prior in light of my own subsequent research, I hope to here provide still greater nuance to the complexity of the paths then sketched, ranging across a wider range of texts from Cixous and Haraway in particular.[5] The transatlantic and interdisciplinary conversation that can now arise between these four figures slips further from the tendency to privilege one as the anchor of the others and more towards a shifting field of relations between humans and other animals that each one enables.

To position the interspecies kiss as performative and thus as a subset of the kiss in general, rather than an aberration, requires a curious kind of work in 2019. While 'performativity' has become part of the lexicon of Visual Culture, often appearing simply in adjectival form (and thus taken to be able to describe a particular form of writing rather than the condition of writing in general), it remains the case that its more complex articulation in the work of Derrida is often overlooked (outside of scholarship specifically engaging with deconstruction).[6] Such is the chasm between what have become disciplinary norms, that the life of the 'performative' in this chapter may be illegible to those schooled in popular readings of Judith Butler or Peggy Phelan.[7] Decades after the work of 'Signature Event Context', Derrida was to issue further caution with regard to the tendency for the performative to continue to foster the illusion of mastering – the illusion that we have the power to cause or to master an event when we are always and already vulnerable to the other.[8] Elsewhere I have tried to show that even when one remains – more or less! – within the realm of what we call language that matters of force, repetition, irony or the countersignature of the other lead us perforce beyond intentionality and thoroughly trouble assumptions regarding the sovereignty of subjectivity up to and including the subject of confession.[9] Writing then in light of both Paul de Man's work on confession and Derrida's critical relationship to it as well as Cornelia Vismann's media theory (drawn upon in Chapter 6 of this book), I found leverage in Derrida's term 'machine-event'.[10] Deriving from his extensive essay 'Typewriter Ribbon: Limited Ink 2' this awkward term did not sublate repetition with spontaneity or vice versa to forge a new concept but brought them into proximity with a hyphen. The mechanicity of repetition and the spontaneity of the event should be categorically distinct. If 'one day' they were able to be thought together as 'one and the same concept', Derrida tells us, this could not even be anticipated by the term 'monster'.[11] It would be so new, he remarked, that it would be 'the first possible event, because im-possible'. Given as a lecture only a couple of years after 'The Animal That Therefore I Am', 'Typewriter Ribbon' too, then, works the ostensible separation between reaction (repetition) and response (spontaneous event) so intrinsic to that now celebrated essay, even as the animal asides of the latter were buried deep in its closing section.[12] It was clear, however, that this contagion deeply implicated the performative: the impetus was 'to think both the machine and the performative event together' because they 'remain ... a monstrosity to come, an impossible event'.[13] In retaining the name of my old friend, the 'performative', here I hope to make clear how embedded Derrida's terms are in each other (rather than naming a discrete array of tools) and to welcome resonance in the thought of others – specifically, Haraway, Cixous, Schneemann. Performativity was already a part of Derrida's subtle displacement of the signifier by means of the text and of the trace; it already impacted on what we thought was human property: rather than the performative providing the vehicle for meaning as that which resides within our organic, present possession, it becomes conditioned by a machinic, reactive repetition (*convention* rather than *intention*). The latter quality was, of course, that which Descartes attempted to section off as the impoverished domain of the animal. It is Descartes' legacy that absurdly fenced off all those myriad non-humans on the side of the machine, dampening down any non-human responsivity.[14]

Insisting that organic and machinic are already within an enmeshed relation inevitably brings Haraway's figure of the cyborg to mind, while the 'doing' over 'being' quality of the performative resonates with her emphasis on the relating of our encounters. While Haraway's contemporary and colleague at the University of California, Karen Barad, published her own arguments for the 'posthumanist' nature of performativity in the same year that Haraway's *Companion Species Manifesto* caught academic attention, albeit drawn out of quantum physics, rather than the troping of biology, the core texts engaged here may be understood as anticipatory of what are now prominent debates on posthumanism.[15] Nervous as Haraway has recently been about the term 'posthumanism' and her own habituation to it, I would suggest two things. Firstly, that her rightful hesitation might be directed more appropriately to the 'post-human' when construed to indicate a temporal break with 'the human' (and the concomitant assumption that we can accurately account for that proper name) and a brighter, even shinier and more cyborgian future that transcends the fallibilities of our assumed-to-be former state.[16] This distinctly irony-free vision takes no account of the decentring of humanism that posthumanism and the 'posthumanities' might more hospitably be taken to solicit. Secondly, I want to resound our editorial comments on Haraway's recent insistence on the earthiness of the compost. In her most recent books, Haraway writes: 'We are compost, not posthuman; we inhabit the humusities, not the humanities.'[17] Pushing the continuities between companion species and compost in the introduction to *The Edinburgh Companion to Animal Studies*, we wrote, 'The humour, even nerdiness, of the figure is an only partly tongue-in-cheek affirmation of the compost with which, and in which, living things compose and decompose.'[18] Moreover, '[g]iven Haraway's insistence on the with-ness of the companion as *com-panis*, we might wilfully hear *com* and *post* as the spatial and temporal markers "with" and "after".'[19] As ever, this is no 'return to nature' but a move that attends to temporal and spatial discontinuities, as such her work proves hospitable to that of Derrida.

The closing sentence of 'The Animal That Therefore I Am' reads '[b]ut as for me, who am I (following)?'[20] It is a citation, wrenched out of context as all citations are and set, unsettled, in another. It is unattributed. Those gathered to hear this essay in its initial form as a public address would doubtless have already heard the traces of Descartes given an afterlife from Derrida's very title and *passim*. Among those gathered to speak at the 1997 colloquium at Cerisy-la-Salle, held in Derrida's honour, under the title *L'Animal Autobiographique*, were Mireille Calle-Gruber, Akira Lippit, Laurent Milesi, Jean-Luc Nancy and Nicholas Royle. Often speaking in the second person, Derrida made quite sure that his address was to 'you', and to 'your' response. The subsequently and posthumously published book of the same title – *The Animal That Therefore I Am* – contained further material and its second chapter, bearing the name '[b]ut as for me, who am I (following)?' immediately treated the citation for the suffering '*suis*' that surfaced within Descartes' *Meditations* in their French translation. The uncertainty that Descartes entertained arises in light of his own supposition of a 'malicious deceiver' out to trick him(self) – the infamous deceiver appears in the second clause of the sentence immediately following the citation.[21] But Derrida opens the disorientation within the present tense itself; its very being is shaken. Throughout *The Animal That Therefore I Am*,

Derrida plays on the first-person present tense of 'to be' – indistinguishable from 'to follow' in French, thereby making thought think again. A deceptively simple '*Je suis*' performatively installs a non-priority and a non-presence in the heart of the thinking subject, the subject formerly known as 'Cartesian'. This transformation is intensified through the substitution of 'The Animal' for 'I think': the anthropocentric quarantine of 'I think therefore I am' is brought up against its excluded other (excluded but homogenized, and controlled through the reach of the concept). Temporal trouble results from knocking the present tense of 'my' thought off-kilter through the implication that I am *after* something else – what? – the animal: it dispenses with linearity, distributes deferral and dislodges origin.[22] If 'Being' is altered through the activity of following, the animal that therefore I am cannot be certain of its own terrain. Following as a hunter? A follower, not leader? One that is always following after the animal, without overtaking the animal, and without pause?

While they encounter each others' work unevenly – one divergently traced by curiosity, science and who reads whom – there is a hospitality that emerges between the thinking of Haraway and Derrida independently of any deliberate mutual involvement. In contrast, the personal and the poetic friendship between Cixous and Derrida threads throughout their writing, explicitly and elliptically.[23] Surviving him in time (Derrida passed away in 2004), Cixous is frequently read in light of Derrida's work. Yet, as this chapter finds, a feline Cixousian uncanny stalks *The Animal That Therefore I Am*, leaving tracks barely remarked.[24] Moreover, however much Derrida may have shifted relations between what we call philosophy and what we call literature (since what he has called 'the law of genre' affirms that they shall always be mixed[25]), his work is perhaps more readily digested as 'argument'. Cixous's thought irreducibly turns on the poetic, and this has long involved animal figures – without subsuming animals under the decorative glaze of 'mere' figures. Observing 'Animals are becoming more and more important in my books' as early as 1996, Cixous invokes the *animot* as early as 1976.[26] The temporal quandary of who I am (following) is presaged in her work when she touches the touch of 'the cat whose cat I am'.[27]

It is unlikely that Schneemann is a reader of Derrida. However, she may well have encountered 'French Philosophers' as compulsory reading at some juncture in Anglo-American art cultures of the 1980s, following on from the lessons in ideology that she parodies with droll delight in her most well-known performance work *Interior Scroll* (1975). In that work, Schneemann famously pulled a scroll out of her vagina, unfurled it and read aloud a script addressing the distaste with which a 'structuralist film-maker' regards her work, detailing its 'personal clutter', 'persistence of feelings' and 'hand-touch sensibility' (a polemic that is actually very funny).[28] The 'structuralist film-maker' stands for that assimilation of the unconscious to ideology prevalent in much theory of the 1970s.[29] That version of Freud and Marx promoted an obligatory critical distance in the arts – along with a widespread ban on the body – that was supposed to disarm the apparatus that would otherwise continue to dupe unwitting spectators into desiring against their own interests.[30] It also militated parricidically against any trace of authorship – or worse, autobiography – as another effect of ideology. Schneemann's work does not enter this chapter as a combatant on behalf of the real lives imagined to be ignored as they wander 'outside-of-the-text' (conceived according to an absurdly literal architecture) nor as a sponsor of the conceit of authorship. Rather, in light of Derrida, Cixous and Haraway, her work indicates a revised sense of the autobiographical, of the touch of the hand, even of what we call experience linking with performativity beyond the speech act and the presumed ability to intend.[31] These revised senses precisely impact on the powers assumed to be human.

The 'autobiographical animals' at play in Haraway, Cixous and Schneemann overlap through the kiss. In the opening paragraphs of *The Companion Species Manifesto*, Haraway writes of one of her dogs:

> *We have had forbidden conversation; we have had oral intercourse; we are bound in telling story upon story with nothing but the facts. We are training each other in acts of communication we barely understand. We are, constitutively, companion species.*[32]

In an essay concerning the concept of the event, Cixous writes, 'With a firm tread the cat climbs onto the woman's lap, looks the woman in the eye with a clear and decided gaze and abruptly a kiss on the alarmed mouth.'[33] In the decades long before the vacant glaze of the smartphone 'selfie', to which we are now habituated, Schneemann, without consideration for finesse, frame or focus, made daily photographic portraits of her morning kiss with her cats, first Cluny and later Vesper, culminating in two large sequences: *Infinity Kisses I* (1981–1988) and

Infinity Kisses II (1990–1998), later made into a short film *Infinity Kisses – the Movie* (2008). The kisses are not identical in all these instances, but they cannot be exhausted by a 'Beauty and the Beast' allegory in which the kiss reveals the beast to have been a man all along. Rather, the kiss solicits an undecidability into the divisions between subjects and species once thought to be decisive.

Kiss Me Honey Honey Kiss Me.[34]

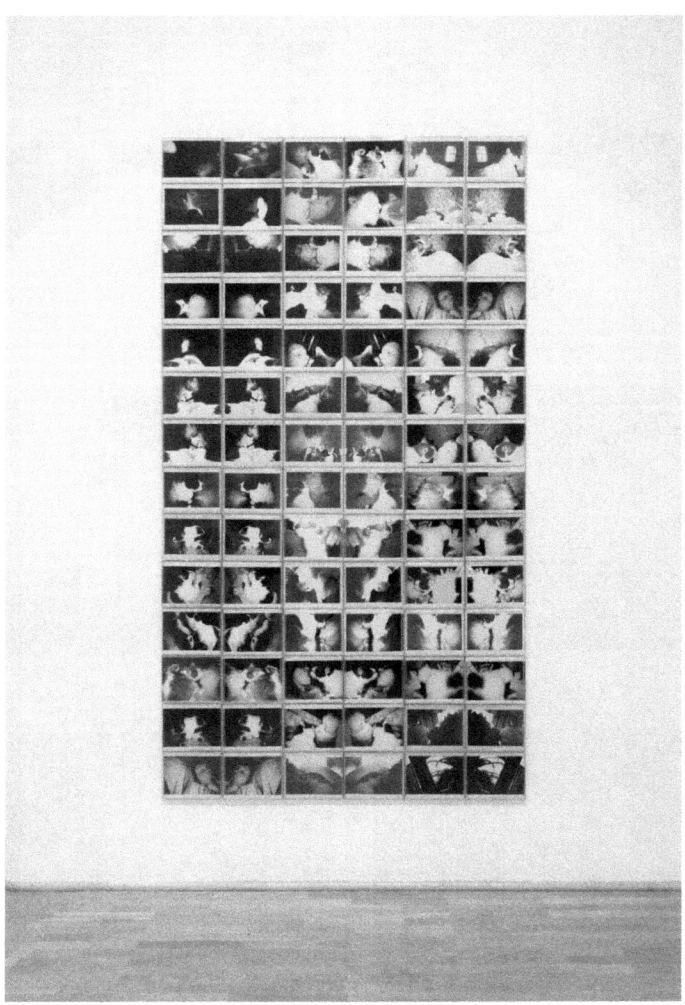

Figure 5.1 Carolee Schneemann *Infinity Kisses I*, 1981–87. Courtesy of the Estate of Carolee Schneemann, Galerie Lelong & Co., Hales Gallery, and P•P•O•W, New York © Carolee Schneemann.

The image that drew my attention to the *Infinity Kisses* is the one that was then shown out of sequence on Schneemann's website from what is a substantial body of images.[35] In contrast to the sexuality bypass affected by most critics of the *Companion Species Manifesto*, Schneemann's image falls most easily, rapidly and ecstatically into one of a grand passion. The woman's head is thrown backwards and reaches to one side in a gesture readily associated with female sexual pleasure (this is not an air kiss; this is not a peck on the cheek). She is

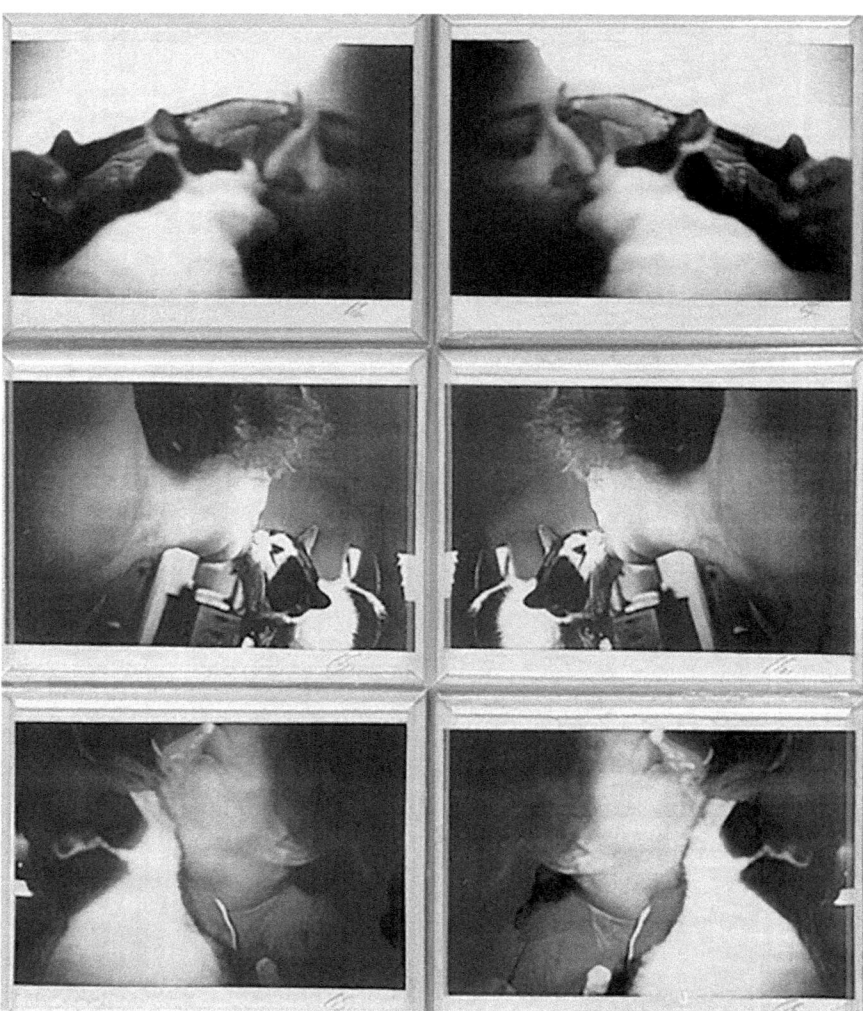

Figure 5.2 Carolee Schneemann *Infinity Kisses I*, detail.

in frame from the shoulders upwards, and since her shoulders are uncovered, her nakedness is metonymically implied. Though the cat is smaller, of course, this frame allows for equal space to be given to both beings. The cat at top right, the woman at bottom left. The cat, perched on a pile of boxes, reaches down to the woman. Their kiss is central: the even expanse of the woman's upper chest and the glide of her neck swoop straight up to it. Breaking up the flow of the woman's neck and the consistency of her skin, the black-and-white markings of the cat's fur call a halt to the traversal of the gaze. Though the woman's eyes are so turned towards the cat that they cannot be seen, we can see the face of the cat, and the cat's eyes are shut. It would be recognizable as a portrait of a kiss even if the title did not name kisses as such. Familiar romantic conventions are out in force; all but one. And this exception, this unfamiliar 'familiar' is captivating.

Infinity Kisses is not a single image, nor a single idea work since the kiss constellates so very much. Rather it is the name of at least two extended series, sometimes rearranged into other patterns, and a movie. *Infinity Kisses I* consists of a photogrid of 140 small 35 mm Xerachrome photographs, each printed twice (right to left and left to right, exacerbating the mirror-image effect), taken from 1981 until 1987, massed together and mounted on linen. *Infinity Kisses II* comprises 24 self-shot 35 mm colour photographs printed as laser images, each 96 × 120 inches, taken between 1990 and 1998. There are strong formal and formalizing elements in the presentation of the work, though the images arise out of the highly informal filming environment of keeping a camera ready to hand in anticipation of the morning kiss that Schneemann habitually received, first from Cluny and later Vesper. Both sets ended arbitrarily, not through an aesthetic decision to render closure or because they work towards a known end but due to the death of the cat (the artist saw Vesper as the reincarnation of Cluny).[36]

Significant art-historical attention has been directed to Schneemann's work and its engagements with space, body, process and ritual – albeit with the tendency to truncate more than forty years of varied practice to a kind of 'greatest hits' list consisting of the individual and group naked performance works, *Meat Joy* (1964), *Up to and Including Her Limits* (1973–76), *Interior Scroll* and *Eye/Body* (1963), as well as the erotic film *Fuses* (1965), for which the artist filmed herself having sex with her then partner James Tenney (in the company of her cat Kitsch, staging an intimate and proximate gaze).[37] This hit list's conversation with feminist theory and art history converges around the question of the body – especially the female body, and agency, sexuality, power and experience. The conversation turns into contestation around the same question. Feminist art-historical analysis of the 1970s and 1980s that was invested in Lacanian psychoanalysis, as exemplified by the earlier work of Griselda Pollock, found it difficult to see any female body step free of the pernicious 'male gaze' to the extent that only signs for the feminine became legitimate means by which to allude to the vexed field of sexual difference. Any female bodies, whether in performance, film or two-dimensional media, tended to be understood as the index of women who had given in to their own inevitable objectification, incapable of producing the requisite critical distance in the spectator. 'Body art' seemed particularly culpable. The feminist art history of the 1990s, exemplified by Amelia Jones's book titled *Body Art* (1998), both challenged the ideological analysis of the problem of the 'gaze' and appropriate responses to it as well as opened performance work by both male and female artists (Schneemann included) to more nuanced analysis.[38]

Much less has been written, however, about Schneemann's works that foreground the feline (photographs, mixed media installations and films – arguably including *Fuses*), the relative silence around which (in terms of critical writing and exhibition profile) the artist is only too aware.[39] *Breaking Borders* and *Remains to Be Seen*, which comprised a large and intelligently curated retrospective of Schneemann's work held in Toronto and Buffalo in 2007, marked a turning point in this absence by organizing her work not according to the deadweight of chronology but according to three themes: war, erotics and felines.[40] With work in animal studies now widespread, obituaries of Schneemann, who died on 6 March 2019, were no longer able to completely sideline this work.[41]

Rebecca Schneider is one of very few art historians to address the feline content of Schneemann's practice, naming the subjects of bestiality, criminality and the everyday in relation to *Infinity Kisses*, the video installation *Vesper's Stampede to My Holy Mouth* (1992) and *Fuses*. For Schneider these works point to the limits, even irrelevance, of gender in light of bestiality, the expected transgression of which is held in check by the quotidian tone of the work. Since Vesper's gender is 'neither apparent nor important', Schneider sees sexuality fade gender from the frame, curiously allowing Schneemann to remain unmarked.[42] Schneider links what she calls this 'art-bestiality' to a multiplication of sexualities commensurate with Derrida's dream of future sexualities beyond negation and thus beyond opposition.[43] While this link is not unwarranted, it does run rough-shod over the range of conflicting work that the rather odd term 'art-bestiality' might be taken to incorporate, as well as – in composing a book about the 'explicit female body' in performance – effacing possible lines of enquiry opened by the question of the animal. In other words, it does not seriously consider the 'questions of interspecies communication' that Schneemann herself acknowledges.[44]

In his prescient book on the representation of animals in contemporary art, *The Postmodern Animal*, Steve Baker glossed *Infinity Kisses*. Reliant on Schneider, he framed the work as the collapse of art into documentary through the weight of the domestic *mise en scène*.[45] Baker's work does imply, however, an interesting dilemma: realism, ostensibly for the sake of the animal, seems to descend into sentimentality, but unfeeling postmodernity only removes the animal by other means – through the sign without allegiance that is allegory. Schneemann's work, I suggest, helps us understand that this division always breaches.

Now, does a kiss respond or react? 'Does my cat really kiss me?'[46] Can we decide? In his recent investigation of the literary kiss, J. Hillis Miller swiftly asserts the metaphoric capacity of a kiss to substitute for speech (act).[47] I suggest that its metonymic function is equally important. If the lips are proximate to speech, then they serve, or at least attempt to serve, as the anchor tying the speech to its actor: these lips kissed you. But, in light of previous chapters, the metonymy that indicates the 'touchiness' of the kiss, that cannot dispel its equal proximity to ingestion.[48] Curiously, in light of Derrida's extensive writings on the incalculable quality of the performative, Miller positions the kiss of which Derrida writes as one that does take effect, even when sent by the virtual means of telephone call, letter and email (by writing, in other words, by the dangerous supplements that expose the absence at the heart of any postal system), in contrast to Kafka's mourning for written kisses, since these are inevitably destined to feed only ghosts and thus never to reach their 'destinee', so to speak.[49]

If a kiss kisses, does it really do so because invisibly, airlessly it seals in the singular affect as *the* intention of the kisser? Does the mark, the 'x' really stand only for the particular kiss it bodies forth, without capacity for standing for anything else? Might the 'x' instead mark erasure? Nay-saying the singularity of each kiss seems a little churlish. No one but a Judas would tear a kiss into some other non-kissing context, surely, and his kiss was pretended anyway. In a debate with colleagues concerning the vicissitudes of the performative, one provided the example of 'I love you' and questioned who would ever say such a phrase and not mean it. Yet there seems to be no better example of this splintered category. Don't we say, 'I love you' romantically and hope that we mean it, or that the saying of it will make it true? Don't we say, 'I love you' spontaneously and wonder, belatedly, if we really meant it? Say it guiltily in response to the demand for reciprocity? Say it habitually, on automatic pilot? Use it accusingly: you *said* that you loved me! Repetition, the sense of convention invading intention, is once again at work, here dogging the human purchase on response. This repetition invites not just a roominess of meaning, a polysemia of the kiss, but an infinity kiss.[50] In this sense it is the kiss that is modified by infinity, by an infinity of contexts: XOXO. Beyond comprehension, since infinity cannot be presented, the kiss becomes sublime. Kiss of infinity: kiss without end.

Miller also ventures the question of morality, hence the possible immorality of a kiss – although, interestingly, this question only emerges indirectly through a citation used to demonstrate the sheer uncertainty of what this non-universal practice might concern. Prior to asking what it means to kiss in general, Joyce's Stephen Dedalus asks whether it is right or wrong to kiss his mother.[51] This morality, or even legality, is not first of all about consenting to kiss or be kissed but rather its confining its proper usage to the inside edge of a law-bound social world. But since kissing is of the edges – edges that lose track of themselves – this confinement is going to be uneasy. Kissing, Miller further notes, 'by no means differentiates human beings from animals, as philosophers used to believe … Chimpanzees kiss, as do other animals'.[52] Unsupported by any references, Miller assumes the empirical explanation that one can observe animals doing the same thing that humans do (kissing). He does not expand the sense of what might constitute this activity through recourse to the less empirically observable problem of the *performativity* of the kiss (his ostensible subject). Neither does he pursue the question, if kissing is common to at least some rather than just one animal, what happens in the inter-species kiss?

Changing the grounds of the 'animal question' is required if anything more or less than a reactive argument regarding the essential quality of the kiss is to be produced. Taking on the near, though not totally, unified front of philosophy that denies language, and hence the ability to respond, to the animal, with an opposing, defensive and reactive claim that 'of course they have their own languages!', like French or Mandarin, gets us nowhere.[53] Derrida, Cixous, Haraway and Schneeemann turn the tables and question our certainty regarding our own clarity of communication; they transform the terms of the alleged privation of the animal with regard to language.

Art historical convention would concentrate upon marshalling adequate evidence for Schneemann's aesthetic process, influence and development, culminating in the signature of the work. Schneemann is signatory; she takes credit for the work as artwork and she names – and defines – a body of photographs as *Infinity Kisses*. In this traditional view, every meaningful aspect of the work is decided in advance, and by the artist alone. But Schneemann opens herself to the risk of an other signatory, naming herself *as a signed body*, as recipient of the repeated actions of Cluny and Vesper: 'he ritualistically, ardently kisses me on the mouth.'[54] She is impressed upon, marked, by those not of her kind. She has also referred to Kitsch as a camera, since her 'steady focus enabled me to consider her regard as an aperture in motion'.[55] An important reversal of subject-centred maintenance of the signatory comes with deconstruction's insistence on counter-signature, that it is the other that signs, the other that offsets the would-be decisiveness of my decision, my signature, my mark, my kiss. Prior to returning to Schneemann, Haraway and Cixous, the following sections elaborate the grounds by which it can also be argued that the cat signs, refining the question of what it is to be an 'autobiographical animal', what it is to kiss – and what it is to *kithe*.

Kithe and kin

Familiar in the albeit antiquated expression 'kith and kin' signifying 'friends and family', the *Oxford English Dictionary* also lists 'kith' as meaning 'information', specifically regarding rules of etiquette: rules concerning what is proper. Its strict separation from 'kin' is unclear. Kinship could also be seen as governing – at least attempting to govern – the proper. Kith is also linked to the verb 'to kithe'. As a verb – a 'doing' word – to kithe is to proclaim, to show or to confess. How performative is that! Derrida's cat does not kiss him; at least he does not tell us whether she does. He is at pains to keep the story of his cat on the most mundane level, in the effort to curb her figural conscription and to keep the discussion relatively directly focused on showing up the philosophical expectation that the animal can only react, rather than, as he finds to his 'shame', respond. The still burgeoning number of texts commenting on 'The Animal That Therefore I Am (More to Follow)', as well as Derrida's other key text from the Cerisy proceedings, 'And Say the Animal Responded?', have frequently fallen into one of two categories. They either applaud the work of the philosopher, relieved that his contribution clarifies both the extent and the stakes of anthropomorphism as well as its flipside of an absolute divide between the human and the animal (such as in the work of Carol Adams),[56] or they backtrack these essays through the rest of Derrida's work to remind us of his consistent investigation of the 'animal question' (such as in the work of Leonard Lawlor).[57] Haraway shares in the relief but supplements Derrida's analyses with both alternative outcomes and a critical relation to what she takes to be the limits of his response to this cat, as well as its visual scenography.[58] On the other hand, since Haraway's priority is to write from the point 'where species meet' as a place conflicted with natural, cultural, colonial, political narratives that incorporate the philosophical, rather than to start from a domain constitutively philosophical, she sometimes cuts short the range to which Derrida's remarks might implicate. The ground may be muddied through following multiple tracks, but that range needs more detailed elaboration here.

A cluster of inabilities accrues to what is called, in a supremely performative gesture of naming, 'the animal'. These inabilities collude in the expulsion of 'the animal' from the terrain that calls itself that of 'Man'.[59] This name-*calling* procedure also invokes the communicative relation of call and response, and from here, the distinction, if there is one, between, response and reaction.[60] In the history of Western philosophy, it is rare for an inability to be reserved as proper to 'Man', yet when it is (in psychoanalysis), this inability or privation will operate as the initial spur that engineers the condition of his dominance.

In Derrida's two hands this last and exceptional inability will turn out to work quite differently, and this difference will mark a general condition.

The titular essay of *The Animal That Therefore I Am* names Descartes, Kant, Lacan and Levinas as key figures in upholding the once-and-forever line between the human and the animal. Dawne McCance makes plain that this theoretical genealogy also took root in the laboratory in which it was apparently possible for Descartes to oversee vivisection while simultaneously holding to the inability of the animal to suffer.[61] The problem coalesces in particular, and in a particularly influential fashion, in the work of Martin Heidegger. For him 'the animal' has no hand, cannot name, cannot speak, has no relation to death, no relation to the 'as such' of the world – in fact is 'poor in world' – and hence has no relation to *Dasein*. Crucially this contrasts with 'Man's' access to all of the above. In order to signal the animal's substantial deprivation, the hand of 'Man' must be, or rather do, something special. It will become clear, however, that this doing, if it relates to the performative, does so in a far from empirical sense (underlining the misrecognition of performative for performance). Firstly, the hand must be given a clean slate, cut off in kind from what Heidegger refers to as the 'grasping organs' of beasts and divorced from the kinship among species named by evolution. It is not that Man's hand does not grasp – the oddity, even monstrosity of its singular form was remarked by Derrida in his earlier essay on 'Heidegger's Hand' – but that his grasp conveys a specialized range:

> I use the word vocation to recall that [for Heidegger], in its destination (*Bestimmung*), the hand holds onto speaking. This vocation is double but gathered or crossed in the same hand: the vocation to show [*montrer*] or to sign (*zeigen, Zeichen*) and to give or give itself, in a word, the monstrosity [*monstrosité*] *of the gift or of what gives itself*.[62]

Moreover, Heidegger joins the work of the hand (*Handwerk*) with what he calls 'thinking': '[e]very motion of the hand in every one of its work carries itself [*träge sicht*] through the element of thinking.'[63] The handiwork of thinking thus becomes a signature of the human: in having no hand, the animal does not think and cannot sign. Thinking of this handiwork as a signature gives a suggestion of the difference performed by the human hand: this hand may grasp, as does the paw, but what it does significantly is give. This hand does not simply, or empirically, give; rather and above all, *it gives itself*. Giving itself, giving the same, the hand auto-graphs. In so doing Heidegger's hand transcends biology (flatly reduced) and accesses the 'as such' of things. The 'as such' of things grants their essence, which remains untouched by the paw or any other 'prehensile organ'. Kithe me.

With so many definite articles invoked, following 'the animal' through the pages of *The Animal That Therefore I Am* is a readerly exercise worth taking literally (along with tracking use of the conditional).[64] Follow 'the animal' and you will follow our bitter conceptual inheritance. Consider the rare use of the indefinite article as a step towards opening the 'corral' of this inheritance.[65] In the more improvisational style of the second essay in *The Animal That Therefore I Am*, Derrida muses at length on his dream of animals dreaming and adjusts his own phrasing and framing. He shifts from condemning the foreclosure in the question 'Does the animal dream?' – foreclosed by the opposition the animal versus the human, reaction versus response – to imagining, as if he 'were dreaming … in all innocence, of *an animal* that doesn't intend harm to *the animal*'.[66] In innocence – that is to say, without the 'anthropo-theomorphic' fall from grace into a state of sin organized by absolute division.

Thus, rather than simply reacting to Heidegger's division of labour by elevating *the* grasping paw, Derrida switches attention to the assumed qualitative separation of giving and taking: this is an element of what cannot be vouchsafed in the doings of any hand. Likewise, the difference between the sign-making capacities of the animal and the linguistic abilities of Man – especially the ability to name, rather than be named – is raised to a higher plane through Heidegger's assumption that the animal has no access to phenomenality as such and thus 'does not unveil the being of being' through naming.[67] Again, for Derrida, rather than reacting and claiming equal footing with regard to being for the animal, what must be questioned is the alleged phenomenal presence in the name of that which is named through showing up its distance as a consequence of the gesture of pointing. Kithe me. Rather than performing a kind of direct capture, pointing is a spacing that disseminates; it is a gesture that cannot gather the 'as such' of that which is pointed out. This privation holds true for the gesture when performed by the hand, even – perhaps crucially – in a relation of auto-affection. The culmination of the animal's privation of hand, language, naming and access to the 'as such' of being is Heidegger's well-known determination that the 'animal is poor in world'. This is in contrast both to the stone, which is 'worldless', and of course 'Man', who is 'world-forming'.[68] In spite of any apparent comparison, Lawlor points out that this insistence on poverty of world should not be read as placing the animal and human on a sliding scale of relation to world. Rather, this is a difference 'not of degree but of essence'.[69] The 'world-forming' capacity of Man again assumes access to the 'as such' of beings. With such an access comes the capacity to question being as such. This ability to question one's own being echoes the gesture of pointing involved in naming, here directed, directly, auto-affectively, to oneself. Never pointing at themselves, never referring to their own being, animals do not die for Heidegger; they merely 'perish'.[70] But, now that the difference between giving and taking, the ability of pointing to gather the world and the presence of being as such in the name are all uncertain, the implication is that Man too does not have access to death as such. If the gesture misleads and Man cannot access, cannot catch up with what – should he do so – would vanquish all possibility (death), then Man too suffers a privation similar to that of the animal. Consequently that common privation muddies the grounds of their ostensible difference. Following this, and since for Heidegger 'the possibility of death defines what most belongs to *Dasein*', Derrida's assertion that these questions shake the foundations of Heidegger's entire ontology does not seem so dramatic.[71]

These unsettling manoeuvres characteristic of deconstruction dismantle the anthropomorphic traps that both identity politics and the logic of 'animal rights' fall into – that is, into defending animals on the basis of the privileges (whether laws or rights or concepts of identity) formerly accorded solely to 'Man'.[72] This is not dissimilar to the shortfall of a feminism of equality – a feminism that assumes that our task is solely to extend the existing range of rights formerly co-opted by men to newly enfranchised women, rather than the ambitions of a feminism of difference that would also examine the conditions and stakes of these rights together with the concept of subjectivity that they assume. Like the latter, deconstruction insists on changing the terms of the argument, such that the grounds of comparison themselves qualitatively change. Neither conceding to the tendency to police an absolute divide of difference between animal and human nor collapsing all distinctions, all signatures, into absolute continuity or identity, Derrida rather refigures the relation between animals and humans as one of 'staggered analogy'.[73] With the stagger comes a spacing: ground not covered by the equivalence marshalled when *this* is corralled into being *like that*. Derrida loosens the assumed anchorage of one side of an analogy. Lawlor presents the staggered analogy as the notion of faults, or differences, without a fall – that is without a Fall from grace, from an original, sinless, perfect, condition of presence. Rather, the staggered analogy opens spacing on all sides, or better, on every shore: for every *corps*.

Im/modest witnesses

Noting that Heidegger does not cite any source to uphold his doxa regarding 'the animal', Derrida pinpoints the discursive pattern to the two major missed encounters in representations of animals: that of the philosopher or the scientific observer of animals, who nevertheless does not see animals as beings who look back, and that of the poets or prophets, 'who admit taking upon themselves the address of an animal that addresses them', but claim to have found 'no statutory representative' who acknowledges the address of an animal in a theoretical, philosophical, juridical or civic mode (other than himself, faced by his cat).[74] Yet, what if not all 'Western human workers with animals' have refused or ignored the respective gazes of animals, as Haraway demonstrates.[75] Turning to (and therefore impacting on) a different archive than that of Derrida, Haraway calls on a range of less discursively regimented work from the sciences that adapt or abandon the flawed concept of the neutral observer (or modest witness). Bioanthropologist Barbara Smuts, for example, finds that her observation of a baboon colony can only proceed when she gives up the attempt to be invisible – better put as hostile – and learns to adjust her behaviours to those of the baboons who certainly are taking note of her.[76]

While Haraway appreciates the mundane quality of Derrida's encounter with his cat, his vigorous efforts to maintain the daily particularity of this cat and to hold at bay the allegorical lure that would dissolve her catness, as well as his recognition that the cat was responding and not only reacting to him, she yet finds a failure of curiosity in his encounter. For Haraway, so much effort goes into addressing the pitfalls that philosophy has entrenched in the discourse of species that when it comes to this cat, if Derrida is curious about what her response might entail, he can find no way to write about it, and she vanishes from his text as surely as if she were a Cheshire Cat. This is one reason I turn to the heightened terrain of the interspecies kiss. The site of curiosity, and the way in which it affects the subject in its grip, is pressing, not least because Haraway has found herself, trained biologist and interdisciplinary critic of technoscience, in the thick of what might be construed as fascist aesthetics: the terrain of 'pure-bred' dogs. Those scare quotes are Haraway's. The second paragraph of *The Companion Species Manifesto* immediately calls up the violent fictions of racial purity that the idea of the breed covets saying, 'One of us, product of a vast genetic mixture, is called "pure bred." One of us, equally product of a vast mixture is called "white".'[77] She has learnt to love playing a sport with – rather than simply maintaining the breed history (herding livestock) of – Australian Shepherds. However, loving Cayenne as a companion animal and as an Australian Shepherd leads Haraway not simply to the closed book of an aesthetics and hierarchy of the type but to another involvement with gene politics. The macro-level leads her to the amateur practices of activist dog-breeders to quash in-breeding (since not everyone follows popular 'sires' for their looks alone, regardless of genetic proximity to their mate) and work towards eliminating genetic disorders through open registries recording pedigree details in full. The micro-level generates the interspecies kiss.

Alternating between poetically attuned diaristic extracts from what she names as 'Notes of a Sportswriter's Daughter' and a comparatively more straightforward theoretical address, Haraway's *Companion Species Manifesto* begins somewhat provocatively. The very first line tells us that 'Ms Cayenne Pepper continues to colonize all [Haraway's] cells – a sure case of what the biologist Lynn Margulis calls symbiogenesis'.[78] (Symbiogenesis refers to new ways of thinking about genetic transfers that do not necessarily involve linear sexual reproduction but rather notice the various exchanges, mergers and recombinations of genetic material perpetrated by bacteria.) Starting off on the other foot, it is Donna Haraway that is 'colonized' by Cayenne Pepper, by the saliva from her 'darter-tongue kisses'.[79] It is the tongue of the other that signs, delivering the saliva that troubles the species of all kisses.[80] This figure is biological in a dynamic sense. It is not merely a 'trope'. Indeed Thyrza Nichols Goodeve has recently given precise attention to this rich term both literary *and* biological: 'a tropism is an obligatory movement made by an organism in response directly proportional to a physical stimulus.'[81] It displaces earlier feminist fears that biology would only ever cut destiny to the whim of those in power. Indeed, the whole of Haraway's work may be understood as a demonstration that, while the biological may be a field riven by power, this field is not therefore forever anterior to culture or otherwise inaccessible, immovable or timeless. Her still paradigmatic early essay '"Gender" for a Marxist Dictionary' historicizes the political urgency of the relatively recent term 'gender' to compensate for the 'deficiencies' blamed on the immutable 'sex differences' of girls, while opening second-wave feminism's foundational 'sex/gender' divide to other differences masked by the genre of gender, as well as, crucially, refusing to give up theoretically on 'sex' or 'nature'.[82]

'Ms Cayenne Pepper' presents a flash of the irony familiar from the first lines of 'A Cyborg Manifesto'. That marital blank 'Ms', rather than the anachronistic 'Miss' or 'Mrs', might initiate the scent of the spectral terror of monstrous lesbian reproducibility. And if there is terror, it may only be compounded by the line-by-line realization that Cayenne is canine, and she is not transformed into a man by the humanizing virtue of the kiss. They have had 'forbidden conversation'; they have had 'oral intercourse'.[83] The wet medium of Cayenne's rich saliva produces a figure of co-constitution and of reproduction, exceeding both species as transfections pass – or communicate – through viral vectors. In so figuring unlicensed reproducibility and in so figuring co-constituting contact as unplanned occurrence rather than appointed moment within the dialectical narrative foretold about the human subject, Haraway bypasses the necessity to surmount any dead fathers in the *telos* of the law. This is not the de-humanization of a sadistic pornographic standard (bestiality) but the de-humanism of life.

Such a de-humanism could have taken hold after the three wounds inflicted on the 'fantasy of human exceptionalism' named as such by Freud – namely, the Copernican wound that decentred the Earth and fundamentally opened the door to 'a universe of inhumane, nonteleological times and spaces'; the Darwinian wound that earthed 'Man' as one amongst other animals rather than their anointed ruler or final outcome; and the Freudian wound itself, wielding the concept of the unconscious to dislodge reason from within.[84] To these Haraway adds a fourth wound, the 'cyborgian, which infolds organic and technological flesh and so melds that Great Divide as well'.[85] These dissolves, however, do not amount to a newfound pool of sameness but a series of differences unanchored by origin or end in sets of shifting relations. These shifting relations need to be re-emphasized in light of Haraway's cited 60 per cent of Americans currently doubting or outright rejecting that humans are descended from other animals.

Play, and all the incalculable risks it entails, surfaces theoretically when Haraway changes the register of the question Derrida uses to reset *our* relation to *them*, and to discontinue any further variation on the rhetorical theme of 'can they do what we (think we can) do?' – the line that Heidegger's work entrenched. Citing utilitarian philosopher Jeremy Bentham, Derrida asks, 'can they suffer?' while Haraway finds more 'promise' in the question 'can they play?'[86] Yet 'can they suffer?' does not simply replace something like 'can they speak? (No? Oh well, let's kill them)'. Rather, '[o]nce its protocol is established', Derrida tells us, implying a direction other than that already established by Bentham, only then will 'the form of this question change[s] everything'.[87] If the question 'can they suffer?' is to exceed a calculus of suffering, it is by virtue of exposure to Derrida's reversal of the philosophical refrain of a privation or lack of an ability as that which is assumed to contrast the proper possession of that ability as mark of the human. This 'animal question' works the ground between privation and ability. This can only spell trouble for the sovereignty of power as *pouvoir* (ability) – they *can* suffer; they are *able to* suffer: '[t]he question is disturbed by a certain passivity'.[88] This certain kind of passivity is a common ground, a common vulnerability. Thus the grounds of Derrida and Haraway do not quite square, but they should be read together, staggering, supplementing each other's work, especially since play too might also trouble the sovereignty of power. If '[p]lay makes an opening', as Haraway suggests, it cannot square the outcome.[89]

While Haraway's 'play' arises in the contexts of her agility work with dogs, in which species must derive a means of communication in order to work together, and of the non-reproductive sexual play observed between dogs in what is the closing sequence of the *Companion Species Manifesto*, 'play' in its more Derridean sense of supplementarity joins the scene at the point where Haraway's interspecies contact zones are also figured as kin-making.[90] With a less melancholic air than 'suffering', those contact zones become labile. I fancy that what we might call a state of 'lability' training would appeal to her sense of what happens *When Species Meet*.

Pet subjects

Haraway's wider argument in *The Companion Species Manifesto* was at pains to extend 'companion species' beyond the 'companion animals' that we think we know under the name of 'pets' (a name that arguably Oedipalizes the animals in its grasp).[91] That wider field should constantly hover even as those domesticated animals with whom we may more easily recognize our intimacy are foregrounded here. Even as the *Manifesto* was nourished by loving Cayenne, Haraway rudely pointed out that her 'red merle Australian Shepherd's quick and lithe tongue has swabbed the tissues of [her] tonsils, with all their eager immune system receptors', thereby conjuring the work between bacterial intimates and mammalian hosts.[92] With her early prompt in mind, we can understand the showy opening kiss of the *Manifesto* to kithe unseen companions. It is crucial, however, for Haraway that companion animals such as dogs are not apprehended as 'furry children … They are not a projection, nor the realization of an intention, nor the *telos* of anything'.[93] She thus refuses both broader sedimented narratives of the domestication of animals as the one-way traffic in their technological appropriation by Man *and* domestic practices that may well name the dog or the cat as the 'family pet' or as a 'family member' (as kin in Shell's terms) but in so doing disregard and disrespect the non-human needs of those animals. Of course, we might also wonder about how it is that we raise humans in our cultures, and what needs we foster, inculcate or render abject, particularly within the Oedipalizing cultures of the West.

Inevitably the (frequently hostile) discourse on pets has informed this research – a hostility also, and typically, levered towards the women in proximity to those pets in their metonymy of domesticity (something Schneeman's art work fought against). Marc Shell's important essay on 'The Family Pet' deserves attention here because it attends to the way in which animals become figured as kin to those not their kind, enabled by the legacy of structuralism and the assumption of a universal incest taboo. Published in 1986, the essay was written without any engagement from available animal studies resources (which could have been drawn out of ecofeminism or rights-based utilitarian thought at that time) and without the questions directed to the embedding of heterosexuality in and as the law that queer theory was soon to muster.[94] That said, Shell did have the insight to question the way that Freudian psychoanalysis failed to notice 'the institution of pethood', even as it entertained animal phobias (discussed in Chapter 1 of this book).[95] Those who defend 'pet-love' in his archive do so for the kinds of sentimental reasons for which Haraway is rightly critical: they are construed to be for us and like us – *our* Family is The Family (this analogy cannot afford to stagger).[96] For Shell, pet-lovers who name their love as one which encompasses all animals fall into a universalism coincident with a Christian ideology of one family, in which 'all men are my brothers'.[97] While this may not immediately sound problematic, it ultimately totalizes that Family. As such, the One Family produces totalitarianism since all occupy the same position; thus, if we are all brothers under God and show kindness to our brothers, those who are not cast as a brother may fear for their lives. Indeed, Shell suggests that this Pauline universalism may turn out to mean 'Only my brothers are human, all others are animals'.[98] In this instance animals are wholly outside the covenant of the law, along with those *called* animals (Shell's caution here bears comparison with that of Derrida in the 'Eating Well' interview discussed in Chapter 2 of this book). Sometimes universalism – as with St. Francis, and as in pet love, Shell implies – incorporates animals as my brothers, my kind. The structural conundrum this now produces is a newfound equivalence between incest and bestiality. Consequently Shell suggests that the logical outcome is 'either universal celibacy and starvation [because *all* are kin, taboos on sex are absolute and likewise nothing can be eaten] or bestiality/incest and cannibalism [all foundations of law fail]'.[99]

While Judaic particularism fares better, in that it does not imagine 'all men' as brothers, thus allowing for some non-brothers to nevertheless be human (and thus does not open a road to the totalitarian) and also renders differences between animals through the laws on Kashrut (thus allowing for food), Shell nevertheless finds it difficult to envision any new kin relation between those we call human and those we call animals. Indeed with the law regarding animals given in Kashrut, he does not pursue why these animals and not others should be eaten. Persuasive as his classical argumentation is, Shell's implicit habituation to the grid-locked relations of structuralist anthropology means that he does not pause to consider if there is any performative contingency involved in the functioning of 'family'. Synchronic slices of current sets of relations (structuralism's remit) just cannot imagine what they (claim only to) describe unfolding in any other direction. Here we should remember the deconstruction of the distinction between description and prescription that goes hand in hand with that between constative and performative.[100] For Shell, the addition of animals makes no operative difference for existing kinship structures: along with a universal incest taboo, there is also an implicit bestiality taboo. No supplement, animals merely compound existing problems.

What we might call Haraway's post-structuralism of kinship assumes otherwise: if humans and other animals are drawn into making new kinds of kin relation, this does not necessarily revert to an Oedipal contract. Looking at the trajectory of her work over several decades shows Haraway's commitment to shattering 'the family' – whether the heteronormative Oedipal nuclear family, the place where eugenics shades sexual hygiene into its poisonous sibling, racial 'purity' in the Family of Man, or in the conscription of companion animals to pethood conceived as being the same as the rest of the family (as in Shell's universalism). In 1997, in the closing pages of *Modest_Witness*, Haraway's heartfelt polemic declared that she was, rightly,

> sick to death of bonding through kinship and 'the family', and [she] longs for models of solidarity and human unity and difference rooted in friendship, work, partially shared purposes, intractable collective pain, inescapable mortality and persistent hope. It is time to theorize an 'unfamiliar' unconscious, a different primal scene, where everything does not stem from the dramas of identity and reproduction. Ties through blood – including blood recast in the coin of genes and information – have been bloody enough already.[101]

While the texts of psychoanalysis are rare in her work, this 'unfamiliar' unconscious knowingly calls out to that which is not of 'the family', even as the logic of the uncanny yokes the familiar and unfamiliar together. Haraway refuses a kin relation that authorizes the pet as (like) a child, since she does not want to infantilize dogs and also because neither 'children' nor brothers exhaust the term 'kin'.[102] Echoing his earlier work on the non-admission of 'sisters' to the category of 'friendship', philosophically conceived, Derrida asks, 'what happens to the fraternity of brothers when an animal enters the scene'.[103] '[T]rying to live different tropes' in Haraway does not retreat into her own personal whims or freshly cleared ground to defend, but it does assume that if there are kinship structures, they neither exist in an everlasting present nor repeat with the exactitude of an idealized machine, nor obey an inexorable law of necessary patriarchal form.[104] *Staying with the Trouble* – Haraway's most recent book at the time this chapter was revised – re-addressed matters with a new forceful and succinct slogan:

> the stretch and recomposition of kin are allowed by the fact that all earthlings are kin in the deepest sense, and it is past time to practice better care of kinds-as-assemblages (not species one at a time). Kin is an assembling sort of word...
> So, make kin, not babies![105]

The impetus here draws its full strength from the question trembling at the edge of the climate crisis in which we are now indisputably immersed, namely population: 'babies' is the metonymy of 11 billion people on this Earth by 2100.[106] 'Kin' becomes a wider, creative force for new alliances and nourishing those repressed of our unfamiliar unconscious that we have previously abjected rather than acknowledged. Self-consciously avoiding or directly countering all the hallmarks of eugenics (as the 'top-down' biopolitical discourse of the right mobilized to encourage the 'right' people to breed and prevent the 'wrong' people from doing so), Haraway's book closes with an SF narrative imagining a future in which shifting assemblages of kinds flourish without that flourishing necessarily repeating the God-given mandate of 'Go forth and multiply'.[107] She pinpoints the ethical urgency of modes of attention that can welcome 'kinds-as-assemblages (not species one at a time)'. This can bridge the symbiogenesis evoked by the kiss of Cayenne Pepper, to the symbionts of 'The Camille Stories'. These stories, which form the last chapter of *Staying with the Trouble*, derive from another colloquium at Cerisy-la-Salle (this time in honour of feminist philosopher of science, Isabelle Stengers in 2013, under the title of '*Gestes Speculatifs*'). In the 400 years or so into the future imagined therein, 'The Children of Compost came to see their shared kind as humus, rather than as human or nonhuman'.[108] In so doing the human population alters dramatically, choosing not necessarily to reproduce and where it does so to often 'come into being as symbionts with critters of actively threatened species'.[109] Earlier in this chapter I suggested a kind 'lability training' as a practice that would speak both to Haraway's affective affirmation of play and the 'nonpower at the heart of power' of Derrida.[110] Let the reader also keep in mind the summons from Haraway's 'Cyborg Manifesto', first published in 1985, 'for *pleasure* in the confusion of boundaries and for *responsibility* in their construction'. This is not abjection.

Shell ends 'The Family Pet' with the confirmation that pets occupy a structural function, a typically anthropomorphic one: 'If there were no such beings as pets, we would breed them', he says, since pets tell us who we are.[111] Yet there is an odd loophole which may well prove to be more than an exception: 'sometimes we *really* cannot tell whether a being is essentially human or animal – say when we were children or when we shall become extraterrestrial explorers.'[112] Science fiction is not necessarily off-world; it is here and now.

Catoptrics: The mirror of autobiography

Along with the disappearance of Derrida's cat from 'The Animal That Therefore I Am' – a disappearance hovering between a failure of curiosity and an ethical refusal – Haraway also finds the repetition of a philosophical trope trapping Derrida in a scene dominated by its visuality and punctuated by his naked body.[113] He is naked before his cat. Naked and ashamed. Yet with Haraway's acknowledged Catholicism and the emphasis of Judaeo-Christian thought on the sins of the flesh as original sin, following Derrida's visual emphasis, she loses track of its equivocality in his text, disappointed that this biblical affect stands before philosophy.[114] On the one hand, she's right – this is about Jacques and not his cat, about whom we learn very little – but the way it is about Jacques makes a difference. His nudity is not of the common garden variety. His shame does not appear as a symptom or as the only due response to the ever-increasing litany of violence against animals perpetrated by humans. Derrida's phrasing is purposefully odd. He writes, 'I often ask myself, just to see, *who I am* – and who I am (following) at the moment when, caught naked, in silence, by the gaze of an animal, for example the eyes of a cat.'[115]

Prompted by the eyes of a cat, this gaze is like a kiss, if a kiss is as open as this chapter suggests. This 'just to see' may seek but cannot seize. Its curiosity is not calculative. Just a few lines later Derrida repeats the phrase, repeats it twice more, and in those instances, it turns back towards him, mirroring the first time, now identified with the cat looking at him 'just *to see*'.[116] Later in the essay it is echoed in the scene of biblical naming. There God lets man (without woman) name 'in order to see' what happens. This God has both an 'infinite right of inspection' and also the 'finitude of a god who *doesn't know* what is going to happen to him with language'.[117] Derrida too does not know what is going to happen when he looks 'just to see' into the eyes of this cat, and in so doing asks after himself. Derrida's human being is troubled, not confirmed, by this other. He says as much. He has 'trouble repressing a reflex of shame'.[118] The intensity of this reflex engulfs and can pre-empt the reader from following the cautionary tale in which it figures. This tale is a reflection, is on reflection. At its 'optical center' 'appears' 'nudity'. His follow-up demands additional emphasis: 'about which [nudity] *it is believed* that it is proper to man, that is to say, foreign to animals, naked as they are, or *so it is thought*, without the slightest consciousness of being so'.[119]

Who am I? Such an autobiographical question would point the self out, point for point 'in the present ... and in his totally naked truth'.[120] This indexical emphasis makes a metonymy for the hand, and the hand recalls its signature place in Heidegger (this is the hand – the hand that *can*). Yet for Derrida the incapacity for the human to point to being as such speaks to the vicissitudes of writing, to the indecidability between giving and taking. The autobiographical animal *would* point out the naked truth, would do so, 'if it were possible'. Jacques Derrida, in the very first line of *The Animal That Therefore I Am*, at the commencement of his address and 'In the beginning', 'would like to entrust [him]self to words that, *were it possible, would* be naked'.[121] The reader must actively seek out, must follow, the conditional tense that so marks this text – indeed it is often our surest [!] guide to the departures that Derrida solicits, departures that lead away from both common sense and the conceptual apparatus of metaphysics.[122] Everywhere you look, there is a conditional 'would'. They gather around ideas that one would otherwise think to be sound. Ill at ease with the conceptual inheritance visited upon his flesh in this encounter both banal and singular, Derrida cautions us, 'Man *would* be the only one to have invented a garment to cover his sex. He would be a man only to the extent that he was able to be naked.'[123] This invention then would mean that we 'would therefore have to think shame and technicity together, as the same "subject"'.[124] This would be the case *if* truth were given by Judeo-Christian ontotheology. The implication, then, is that full disclosure, of the nude, of naked words, of the truth as that which bares all has never been within our grasp and never will be so.

Derrida's opening desire to trust himself to 'naked words' – were it possible – might also be read ironically in light of Haraway's own exposure of the illusion of the unmarked subject served by the 'experimental life' of the eighteenth century (an illusion served *to* the history of scientific practice in the West). Drawing on Steven Shapin and Simon Schaffer's book *Leviathan and the Air Pump: Hobbes, Boyle and the Experimental Life*, Haraway recounts that the ideal witness to this life was a modest one, yet one whose modesty must be visible while the witness himself should be invisible: '[t]his self-invisibility is the specifically modern, European, masculine, scientific form of the virtue of modesty.'[125] This manner of invisibility counted on the witness 'adding nothing from his mere opinions, from his biasing embodiment.'[126] While the witnessing (most famously, of an experiment on a bird in an air pump) should take place in public, and taking into account the vexatious determination of just which context and by virtue of what audience would constitute a 'public' event, perhaps we could think of Cerisy-la-Salle as such a forum and Derrida's 'experiment' his reflection on the animal. Moreover, the bad feeling – the shame – that afflicts Derrida and draws attention to his body, naked before his cat, thoroughly troubles the invisibility of his witness. Rather than completely dismiss the figure of the 'modest witness', Haraway seeks to 'queer the elaborately constructed and defended confidence of this civic man of reason in order to enable a more corporeal, inflected and optically dense ... kind of modest witness to matters of fact to emerge.'[127] Might the embarrassment of the philosopher drawing attention to himself and his body in this way in fact speak to the corporeal inflection that Haraway entertains as she puts pressure on the necessarily modest quality of witnessing as partial or even 'situated'?[128]

> Moreover Shapin and Schaffer described Robert Boyle as himself professing a 'naked way of writing':
> He would eschew a 'florid' style; his object was to write 'rather in a philosophical than a rhetorical strain.' This plain, ascetic, unadorned (yet convoluted) style was identified as *functional* ... Moreover, the 'florid' style to be avoided was a hindrance to the clear provision of virtual witness.[129]

The same can hardly be said of Derrida. While not necessarily 'florid', Derrida's style could hardly be described as 'plain' or 'unadorned'. Naked before his cat – 'stark naked' as the translation has it – Derrida is '*á poil*': he is down to his 'animal' hairs, in common with mammalian animality.¹³⁰ Immodest, marked, partial, situated: the philosopher's body responded. That this response speaks also to the beat of the heart will be taken up in my closing chapter.

Derrida does not look in order to annihilate the difference of the other by seeing only his own re-confirmed reflection ('pointing' straight back to him). Rather, the other can always surprise. For where autobiography habitually imports a mirror to figure its reflective function and to lead back automatically, autoaffectively, to the signing self, 'The Animal That Therefore I Am (following)' sees itself in the eyes of a cat. The mirror of autobiography that would see all is not a '*miroir*' but a '*psyché*' in Derrida's text: a full-length mirror that can be angled to reflect what one desires.¹³¹ Habitually, in post-Lacanian thought, it is by means of the reflected other that the subject is staged for itself as such: misrecognition of that subject's capacities is part of the allure. Yet what if the other is wholly other, and not a brother in advance? What if this other happens to arrive in the form of a cat? The autobiographical mirror is like one found in a bedroom, the place in which one undresses. But in this context, it also conjures up the *invention* of the other – the subtitle of an essay and two collections of Derrida's essays: '*Psyché:* the invention of the other'.¹³² Jacques Lacan once opened limited access to the mirror for the animal as a rudimentary visual reflection functionally tied to sexual maturation, but without speculative insight, without the dialectical spur that would technically supersede privation (that privation being the alleged singularity of human infantile immaturity in this case).¹³³ There the recognition of the animal's own image did not forecast the ability to say 'I' as it did for the human infant.¹³⁴ Neither could animals take each other's images for their own. But seeing himself in the eyes of a cat, signed by this other, one morning in the bathroom by chance, Jacques Derrida becomes an autobiographical animal, and he is not alone.

Supplementing the gaze of the non-human other, it is the interruption in and of the kiss that is underlined in Cixous. She does not see it coming, even though her eyes are open and the cat's determination is flagged by its 'clear and decided gaze'.¹³⁵ For that matter she does not see the cat coming, either: 'in the meantime the cat arrived.'¹³⁶ Pure event, and as such, astonishing. Not 'in the beginning' – Derrida's opening clause that announces not his own obeisance to Judaeo-Christian onto-theo-chronology but his shame before the violence of that tradition – but rather, '*in the meantime* the cat arrived'. Cixous is struck by unforeseen hospitality in the meantime. Arriving without an article, as the chapter of her novel *Messie* (1996), '*Arrivée du chat*' stresses arrival even over the cat, in a sense that slightly dissipates in its 2006 English translation as a discrete essay called 'The Cat's Arrival'. Even so, published in a special issue of the journal *parallax* on 'animal beings', this essay has been interjected into a field rippling with the after-effects of Derrida's work on the animal question, but which still struggles to keep pace with that of Cixous. At least two volumes now invite her work to be read in this light: Marta Segarra's edited and annotated collection of her writings *The Portable Cixous*, published in 2010, explicitly names 'The Animal' as a core theme for Cixous (and one not limited to cats or domesticated animals).¹³⁷ My edited collection, *The Animal Question in Deconstruction*, published in 2013, expanded the practice of deconstruction beyond the proper name of Jacques Derrida and opened with Cixous's essay 'A Refugee' as well as including essays that foregrounded this practice as one of poetics as much as a politics.¹³⁸ But Cixous's ceaseless breaching of genre disorients the difference between autobiography (that should only ever point back to the self) and magical realism (that should always escape the perceived realism of the self). Thus the challenge of her prose and her preference for embedding 'argument' within poetic phrasing and neologism may continue to fox those readers hoping to find a programme or a method that they can grasp: as Cixous herself writes, 'I do not command. I do not concept.'¹³⁹ Of course, these cannot be found in Derrida either!

Cixous's cat does not make an appearance only to subsequently fade from view, or atrophy in allegory. But the stress is on the surprise of the encounter: she writes, 'I'd also never have imagined … That the Event would be a cat.'¹⁴⁰ Her surprise at this unforeseen, and thus eventful, remarkable encounter foreshadows that of Derrida: both of them rendered not at home in their own homes, both caught on the unhomely, uncanny, backfoot.¹⁴¹ Perhaps this was 'the first possible event, because im-possible'.¹⁴² In Cixous, since arrival is a feminine noun in French, this force of arrival is stressed as feminine, while the cat of the title is in the generic that is masculine form until the first line of the essay, when 'la chatte' is specified (had the French used 'arriver' as a verb – *Le chat est arrivé* or *La chatte est arrivée* – it would have both emphasized the cat and obliged agreement of the choice of gender across noun and verb).

A similar sense of arrival arriving prior to any assignation surfaces in Derrida in his ambiguously wrought term *l'arrivant* – forgetting, he tells us in a footnote, that Cixous had made use of '*arrivant*' in her novel *La* (1976) and the next year produced a play based on that book called *L'Arrivante*.[143] On the one hand, this footnote frustratingly contributes to the play of precedence between Derrida and Cixous, whose works echo each other so much anyway, not least in the context of deconstruction's troubling of precedence itself – who am I following?[144] The door and the shore ('*la rive*' conjuring '*l'arrive*') marking the arrival of the *arrivant* – repeating throughout Cixous's text – redouble this problem, as Derrida remarks: 'this border will always keep one from discriminating among the figures of the *arrivant*, the dead, and the *revenant* (the ghost, he, she, or that which returns).'[145] At the core of deconstruction then, this *corps* never fully manifests, is never caught, remains unsure. The one who arrives should be unmarked because unforeseen, a neutral if not neuter arrival. On the other hand, this is also fraught in the general context of work by women, so rarely cast as eventfully paradigmatic rather than machinically derivative – and, in dating from the mid-1970s, *L'Arrivante*, like *La Jeune Née*, marks a time when it was most pressing to challenge alleged sexual neutrality (*La Jeune Née* is translated as *The Newly Born Woman* though only the feminine not the species is strictly specified).[146]

Cixous's writing encompasses the literary liberty of free indirect style, enabling rapid shifts between first- and third-person voices, dissolving easy identification or solidity of 'character', and frequently provoking the question 'who speaks?' Thus in her text, both 'she' and 'I' are caught off guard by the arrival of the cat. The question is intensified, given the shift in context, the grafting of '*Arrivée du chat*' from an ambiguously fictional and ambiguously autobiographical book into a journal context in which we expect the first-person singular to be the vehicle for the author's own voice, to sign for the author and for no other. Provoking 'who speaks?' or 'who signs?' usually skews the machinations required for rendering and retaining the human subject as central. Here it enables Cixous to pose and repose all the clichés of anthropomorphism without simply falling into thoughtlessly enunciating them herself. She lets fly the distance of the literary while at the same time appropriating autobiographical authority, saying that she had never thought that a cat would show up, not 'in one of my stories'.[147] Remarked in this way, Cixous's 'stories' become her autobiographical animal, and events exceed her control. Not simply a personal quandary, events exceed human control as they let both animals and machines rush in.[148]

In Cixous this cat that arrives can speak – but in order to demonstrate projection, to reveal error. In the middle of the essay, the narrator elects to bathe the cat installed in the narrator's heart and house despite protestations.[149] This bath, however, is imagined as a denuding disaster in which her fur dissolves away completely, necessitating that the narrator cover her with a towel. Amid the echo of shame as their bodies are made if not identical then similarly in need of the modesty of clothing, Thea reproaches the narrator for 'passing the limit' and names her transgression as the 'exaggera[tion of] love'.[150] In Cixous the ruse that the cat is really and merely an allegory of the displaced human partner is voiced to try and ward off the animal's difference and the possibility that the cat is a rival (as Anglophone readers will hear). When comings and goings have already circumvented his post, that displaced partner – his improper name a literary quotation (Aeschylus) – is reduced to attempting to reinstall proper boundaries. He grasps pointlessly at the functioning of doors, hence laws, and who should remain before them, as a limitless amount of women and fantastic beasts traverse this open house, a hippopotamus highlighting the hyperbole.[151] Meanwhile the woman comes to offer unconditional hospitality to the others to whom she is hostage, in spite of Emmanuel Levinas, for whom the animal has no face and hence no relation to the ethical.[152]

Attempting to refuse to accommodate a foundling cat for any second longer than necessary, the narrator takes this cat – who will become Thea – to her cousins' house and leaves her there. But the cat, with a newly torn ear, returns and not only returns but 'came to thank the hostess for saving her from the hell into which she herself had thrown her'.[153] 'You're thanking me!!?' the woman asks, on the cat's return.[154] Startled by her affirmative kiss, given in disregard of the woman's initial inhospitable shock at their proximity, the woman 'asked herself who could have taught the animal this way of thinking'.[155] So close together in the text, alliteration pulling them together, 'thanking' can easily be taken for 'thinking'. As Cary Wolfe notes, thinking and thanking are etymologically entwined.¹⁵⁶ Thea thinks and thanks, and these things redouble the surprise that she be there at all, challenging anthropocentrism not simply by refusing to project human qualities upon non-human species but questioning whether we can properly identify the *anthropos* as such.¹⁵⁷ No gratuitous slippage, the move implicates Cixous in the same critical relation to Heidegger's '*Handwerk*' that Derrida more explicitly poses. The gift, for which one gives thanks, was supposed to sign for the human. It was not supposed to be performatively available beyond the human, not supposed to be so radically open to citation.

A mirror crops up in '*Arrivée du chat*', again as the device that points back to the signing subject. Pre-figuring Derrida's nude scene, it is an uncomfortable return: it is host to a supplement of shame. The mirror moreover is personified – or 'animalified': the mirror 'squawks' like a parrot, it is 'perched' on her shoulder, it will not shut up and it is repetitious: 'Naked naked naked see you see you see you', exacerbated in the phonetically close French: 'Nus nus nus vus vus vus'.¹⁵⁸ Cixous is naked before a Cartesian-Lacanian animal-machine – the *psyché* that parrots the Judaeo-Christian history of technology inscribing nudity as shame. This parroting mirror that shows will always show you up, land upon the fault. A moment later the mirror derides her in terms that redress her nakedness as a type of clothing, 'a get-up': 'look at that get-up it isn't pretty.'¹⁵⁹ Pointed out, 'in her totally naked truth?' The mirror affronts her. The mirror shames, barring her from the invisible role of modest witness. Its plural derision is also addressed to the cat ('*vous êtes nus*'), but sits on the shoulder of the woman recalling her to its specular economy that always sees the same thing. No insult from the woman's exasperated human family quite manages this. '*Arrivée du chat*' specifically links the inculcation of shame with the technology of the mirror. The mirror is charged with splitting the pair of them into the self-conscious one that knows and must overcome her shameful nakedness – or proximity to animality – and the one that does not. The story of shame is thus embedded within the original sin of speculative dialectics that spurs the human to produce both clothing and the veil of consciousness as a technology that sublates the animal. What if dialectics offers a history of technology both incomplete and idealized, polished up to cover the faultlines? This is the charge that both Cixous and Derrida level at the mirror, the *psyché*.

Animotion

For Schneemann, every morning the awkward analogue camera was there 'just to see', without determination over the image (and without the now-habitual glazed gaze of a selfie captured with the high resolution of the now-ubiquitous camera phone). She had it there, ready to hand, by the bed, ready to turn towards her cats as well as towards herself. Schneemann is in the photographs, she signs: quickly read, and without the insight of the theoretical work drawn upon in this essay, *Infinity Kisses* would be legible as autobiographical work as it is traditionally and simply identified – as an index of that which is 'my own', and legible within an artist's work known for its investment in the personal, in experience, in a 'hand-touch sensibility'. But does that hand point back to the signing subject, and point her out, point for point in her naked truth? 'Nudity perhaps remains untenable', remarks Derrida, his own desire for naked words, for 'words from the heart' having been revealed to have been a *dressage* of sorts.[160] Nudity cannot be held in the hand, gathered in a moment of presence. Schneemann's nudity precedes her: 'nudity' clothes the performance works more readily associated with her name as a form of signature. She is always partly concealed, as are her cats, and this is not a violence that could be avoided according to a more ethical representational strategy that would complete the picture without exclusion. Despite Derrida's caution, he warns readers that we not be able to decide for certain whether his story was true or fictional, since the telling and the reading of that story will always redraw its contours. Texts of whatever context cannot police their own borders.

If the mirroring function within Schneemann's works was perhaps less clear in the first version of *Infinity Kisses*, where the smaller images were abutted against each other and mounted as one totemic grid, in the subsequent versions of much larger prints, the effect is magnified. Glazed, they reflect viewers as well as other prints. In Schneemann, the trope of the mirror is no trap of the same. The eye of the camera is a frequent trope for the 'I' of a directing subject (as Schneemann well knows, given her derailment of this authority in

Figure 5.3 Carolee Schneemann *Infinity Kisses II*, 1990–98, 2004 print edition as shown in *Remains to Be Seen*, CEPA Gallery, Buffalo, NY, 2007. Courtesy Galerie Lallouz

her *Eye/Body* work). Here each blink of the shutter releases an other image – image after image of – what? As Cixous writes, 'Woman with cat? Or Woman belonging to cat? Or Cats? Or Woman? Or Women? Or the foreigner?'[161] Who mirrors whom? *Infinity Kisses* is not reducible to the too-available and pitying cliché of older women clinging to the company of cats as poor substitute for the men they can no longer attract. We must take account of a mirror of technical reproducibility, a mirror that iterates – a point reiterated by the various articulations of these photographs, re-cited in different contexts when Schneemann uses her own work as a found object. In so doing she emphasizes their range beyond realism. Their eyes are shut in most images. Even when they are open, the inability to predict frame or point of focus renders the visual field contingent. Schneemann's experiments in framing – horizontally or vertically, and variations on printing the images in pairs as loose similarities or as precise doubles – makes use of photography's propensity towards infinite reversals between positives and negatives and allows no gaze to resolve a hierarchy ordering who follows whom. *Infinity Kisses* names the work, emphasizing the performative kisses without consolidating their content or contracting their context. The kiss brings to attention the entanglement of response and reaction without dissolving those who kiss into a pool of sameness: kissing is of the edges, of contiguity, not continuity. Performativity is neither a power nor the proper of the human. Kithe me.

6

Outlaws: Towards a posthumanist feminine after *Dancer in the Dark*

'*Tympaniser*', Alan Bass tells us, is an 'archaic verb meaning to ridicule publicly' or to decry.¹ In the essay fronting *Margins of Philosophy* called 'Tympan' (translated by Bass), Derrida decries the philosophy that would own its limits, absorbing 'the margin of its own volume'.² This early attention to the exorbitant staking of territory through self-address – philosophy *calls itself* philosophy – anticipates remarks across his work regarding the address of the one who calls himself 'man' and the other 'the animal', as if this calling could maintain the fail-safe erection of a proper limit.³ It is worse than performative arrogance. The '*relevance*' of the limit, at the beck and call of philosophy, is embedded in the Hegelian sense of overcoming (*aufheben*, sublation, *relever*) that it quietly harbours.⁴ This limit always strives to overcome the other, without leaving a remainder, and thus to limit any incursion. The outside is domesticated to the outside *of* (belonging to) philosophy. This is the speculative philosophy of which Freud denied all connection even as he repeated this gesture, posing the limit as forever '*da*' (to recall the gesture of Chapter 4).

While the limitrophy between species continues to nourish scholarship on deconstruction through multiplying differences rather than maintaining vain adherence to human exceptionalism, 'limitrophy' is also named as the general condition of the interface of the limits in the much earlier text, 'Tympan'.⁵ Not the atrophy of the limit but that undecidable event between nourishment and conquest, thickening and feeding the limit.⁶ The limit impresses 'according to new *types*', it leaves its stamp.⁷ As I have elsewhere argued, for deconstruction '[l]imitrophy becomes the law'.⁸ In 'luxating the tympanum', the limits, or indeed the *margins*, of philosophy are eaten.⁹ Equally, given the percussive address of the tympanum, we might say that the margins of philosophy are beaten (typed upon, stamped upon, caused to vibrate).[10] Thus, when the three epigraphs from Hegel regarding sublation, limit and the proper 'vestibule' of philosophy are situated on the first page of this typographically precise text ('Tympan'), and Derrida's own typing begins on the second, that is to say the reverse of the first page, then, through playing on the structure of thesis and antithesis, we find Derrida beating Hegel on the backside (in contrast to the more typographically conventional remaining chapters of *Margins*).

Struggling within the strictures of a small-town amateur dramatic rehearsal, Selma, the lead protagonist in *Dancer in the Dark*, makes a plea for more rhythm: for 'some drums, or something.'[11] But there is no license for her tap-dancing in *The Sound of Music* (the production that is being adhered to with gruelling literality). This chapter considers *Dancer*'s persistent play on rhythmic sounds as that which generate Selma's fantasy song and dance numbers as a 'tympanising' of the limits, notably of the limits of the law, as they also work the margins of the film's two styles (broadly, realism and musical). In his provocative analysis, Cary Wolfe suggests that we might understand Selma's vocal style (given singular expression by Björk) as a refusal of the phallic imposition of language and that her virtually suicidal submission to the death sentence in the film's denouement, construed as a Lacanian 'feminine' act, evokes the notion of a 'posthumanist feminine.'[12] 'Outlaws' replays Wolfe's tantalizing conjunction according to another type of thought, one more in tune with that of Derrida. It draws on Derrida's late work, in which it is polemically clear that deconstruction has never been confined either to the signifier or to the human, and sounds out the implications of his early work on b/eating the limits through Selma's oblique tympanum.

Earphones

'Tympan' is paginated in roman numerals, thus marking it out as a kind of introduction bearing upon the rest of the essays collected in *Margins of Philosophy*. Along with the typographic intervention into epigraphic conventions already indicated, the body of this text bifurcates into two uneven columns formally enacting the patterns of which it speaks: the broader left-hand column channels Derrida's exegesis; the narrow one on the right is a long citation from Michel Leiris's autobiographical ruminations *Biffures (Scratches)*.[13] Such sensitivity to the page and its architectures is hardly idle: 'Tympan' predates the (French) publication of *Glas* by only two years, the book whose wildly singular structure dramatically offsets a column invoking Hegel and another, Jean Genet.[14] The right hand of 'Tympan' follows the associations of what Leiris calls a '*spiraled* name' that joins earwig – '*perce-oreille*' with Persephone.[15] Derrida focuses philosophical attention on the ear, with its labyrinthine canals, its oblique drum – the tympanum, and its inner hammer. He demonstrates how the ear is imagined to be in complicit resonance with the philosophical and specifically Hegelian understanding of the limit as the proper by 'produc[ing] the effect of proximity'.[16] Hearing then, in this understanding, is again marked out as hand-in-hand with the metaphysics of presence. Crucially Derrida discloses that, Hegel's system of thought notwithstanding, the limit already opens onto 'the stereographic activity of an entirely other ear'.[17] Substituting 'phonic' for 'graphic' does not mean that he militates 'against' sound in some fundamental way (on the contrary) but rather that Derrida emphasizes a writing machine at/as the heart of the ear (where we might expect 'stereophonic', he writes 'stereographic'). Also departing from conceptual expectations, this writing machine can helpfully be approached in light of Derrida's late essay 'Typewriter Ribbon', in which machine and event must become hyphenated if the future might welcome a truly new logic of thought, a new type indeed.[18] This impossible machine-event refuses the opposition of repetition and spontaneity, inorganic and organic, that the two terms *qua* concepts habitually convey.

In a rather downbeat invocation of 'Tympan', Veit Erlmann suggested that the tympanum 'requires more than its being capable of resonance and of casting back philosophy's *logos* on itself'.[19] This downplaying, indeed limitation, of what is hammered out in 'Tympan' is issued on behalf of feminism, indexed for Erlmann by the work of Luce Irigaray. Irigaray may well want a positive theory of the feminine that is both Other than the Same, to use her framing of the negation of the feminine typically produced in metaphysics, *and* other than the destinerrant inconstancy of sexual differences found in Derrida (discussed in Chapter 4 of this volume). However, and as this chapter will demonstrate, it is highly unlikely that Derrida fails to recognize the tired tropes of femininity with which Western philosophy is littered, as Erlmann suggests.[20] Clearly the tympanum must do more than resonate, more than cast back logos on itself, if there is to be any new movement, Derrida argues as much himself. He is at pains to 'ambush' the logos in a manner other than the frontal, oppositional logic that it knows so well (punning on the homonymic Greek *lokhos*, meaning both 'ambush' and 'childbirth').[21] Repeatedly describing his practice as one of luxating the tympanum, Derrida's seemingly odd choice of verb positively avoids one that begins with a negative prefix like 'de-' or 'dis-' (as in 'dislocate'), again remarking his side-stepping of antithesis. Ambushing the logos, he sets the '*loxōs*' [Gk. oblique] to work within it, licensing this oblique with the indirect angle of the tympanum.[22] The logocentrism from which Derrida wishes to take leave is already identified in 'Tympan' as 'phallogocentrism'.[23] He challenges the habitual philosophical resonance between the tympanum and matter as both virginal and feminine, destined only to faithfully 'repercuss' the other's type.[24] Following the gestures of Chapter 4, however, the indiscretion of this iterative ear can never patent an impression of 'her' own either.

In a refreshing solicitation of the concept of the limit that radically departs from a phallogocentric phantasy of inscription as insemination, Derrida gives up the fetish of the supposed 'virginity' of that which is outside the philosophical text and acknowledges 'another text, a weave of differences of forces without any present center of reference'.[25]

Dancer in the *Arkhē*

Dancer in the Dark cast Björk in the lead role as Selma Jezkova, an immigrant in a small American town in 1964. Hiding her impending blindness, Selma saves her hard-earned money to protect her young son from the same fate. Her landlord, the apparently respectable Bill, steals that money. Rather than directly return it, the depressed man begs Selma to kill him, which she does. Her spurious trial results in a death sentence. This chapter turns to *Dancer* in light of, but taking significant leave from, Wolfe's insightful analysis of this film. The divergence is far from a reactive one of questioning any engagement with a 'bad object'. Wolfe first drew my attention to *Dancer* as something potentially more complex than director Lars von Trier's *enfant terrible* reputation might suggest. Von Trier is well known for numerous controversial films, frequently featuring lead female protagonists assailed by traumatic plots (e.g. *Dogville, Antichrist, Melancholia*).²⁶ Rumours at the time of production that von Trier treated Björk badly subsequently came to public attention with the gathering momentum of the 'Me Too' movement, in which women spoke up regarding sexual assault in light of the multiple allegations made against Hollywood film producer Harvey Weinstein. In 2017, Björk publically posted of her experience of harassment from von Trier on the set of *Dancer:* the post remains on her Facebook page.²⁷

Perhaps it was the vicissitudes of publication history that explains Wolfe – incorporating an essay from 2001 in his 2010 book *What Is Posthumanism?* – initially being persuaded by Slavoj Žižek and Mladen Dolar's negation of deconstruction especially on the question of 'the voice'.²⁸ In the intervening decade, by contrast, Wolfe has effectively led the field in drawing nuanced attention to the 'question of the animal' through Derrida's work, including publishing the first English translation of 'And Say the Animal Responded' in the prescient *Zoontologies* collection (Derrida's second major critical essay on Jacques Lacan and the second essay to be published from the Cerisy la Salle colloquium *L'Animal Autobiographique*).²⁹ Most recently, Wolfe both secured the republication of Derrida's *Cinders* and wrote a significant introduction to it – 'Cinders After Biopolitics'.³⁰ There he revisits the status of voice, given the multiple citations from Derrida's previous work that make up the left-hand pages of the book form of *Cinders* (again, a formal experiment pressures the event of reading the book). In *Cinders* the trace is as unavoidable as it is ungraspable.

In 'And Say the Animal Responded', Derrida finds the theory of subjectivity in Lacan's *Ecrits* to repeat the Cartesian division of man from animal as that of response from reaction. Derrida further remarks upon the idealization of signification and the consequently narrow construal of the field of language legislated by Lacanian psychoanalysis and reminds readers of his own insistence upon the trace as that which could not categorically render such a divide. It is thus disconcerting when Wolfe turned to the signifier in an admittedly qualified use of Žižek 's work.³¹ This signifier, naturally, is the phallus. Highly cognisant of the more obvious pitfalls of tarrying with such concepts, Wolfe cautions engagement with Žižek through recourse to the feminist critique of Judith Butler (citing *Bodies that Matter*).³² Consequently, in order to pursue his advocacy of an ethics of the posthumanist feminine, he calms this caution through 'insist[ing] on the phallus as *signifier*'.[33] In spite of her extended arguments with Žižek regarding the phallus, the Real and the contingent, I would argue that Butler, as fellow Hegelian to Žižek, is not supplement enough.[34]

Without rehearsing the entirety of Wolfe's extensive engagement with *Dancer in the Dark*, suffice to say his attention to the cinematically rich and unusual play between two distinct styles in this film leads to an equally rich discussion of competing understandings of the world. Not just any styles, the film alternates between a dirge-like *verité* and more colourful, musical fantasy sequences, albeit a musical offset from a completely familiar form through the casting of Björk. Following these two styles, Wolfe examines the world of appearances apprehended by reason versus direct engagement with 'the thing in itself', as mourned by Stanley Cavell. Framed by this discrepancy, Wolfe's essay seems bound to conclude with the psychoanalytic mainstays of Real and Symbolic. Given that *Dancer* provides so much on which to remark regarding the use of voice and sound more generally and that the film also focuses on the figure of woman before the law, Wolfe broaches his psychoanalytic engagement through the work of Kaja Silverman. While Silverman provided path-breaking attention to the question of the female voice in both Hollywood and feminist counter-cinemas at the time when scholars could only see the gaze and the screen, and her work thus meters Cavell's inattention to female embodiment, she appears here as exemplar of feminist misunderstanding of the phallic hegemony of/in psychoanalysis.[35] Rather than see Selma as the all-too-familiar feminine site for the projection of losses *a lá* Silverman, Wolfe initiates another reading through Žižek, who 'would identify the phallic itself with such losses and would therefore locate "the feminine" at the very core, and as the very truth of, the phallic'.[36] Following Lacan and Žižek, Wolfe then frames the phallus not as the apotheosis of masculine power but as that which is retroactively installed, exposing the fragile point at which the subject is supplemented by the Other. 'Supplement' here is Žižek's term, mobilized in the text in which he asserts the identity of Derrida and Lacan through what he claims to be the equivalence of supplement and signifier, 'Master Signifier' no less. In a single swipe then, Žižek attempts to annihilate *Of Grammatology*, a move that is in step with the Hegelian logic of the limit addressed in the introduction to this chapter. Blindsiding the breadth of forms that might function as a supplement, Žižek seeks to tie *Of Grammatology* to the linguistic terrain from which it otherwise shakes loose.[37] Ironically, Derrida notes that when writing *Of Grammatology*, the discourse that was 'the closest and most deconstructible, the one that was most to be deconstructed was no doubt Lacan's ... with regard to the primacy of the signifier'.[38] Wolfe, however, here accepts Žižek's exposure of the fiction of the subject to mean that the phallus should be 'properly understood' as 'always already posthumanist and, in that sense, "feminine"'.[39]

Now, 'posthumanist' and 'feminine' are terms that are very attractive, and together they are tantalizing indeed, calling to my mind the aside that Derrida makes regarding 'almost all the[se] animals' in his work being 'welcomed ... on the threshold of sexual difference' and thus infringing upon the subject *qua* human and masculine.[40] Here we should recall the way in which the subject *calls himself* 'Man' and that the 'authority and autonomy' thereby endowed are '*attributed* to the man (*homo* and *vir*) rather than to the woman, and to the woman rather than to the animal'.[41] That 'posthumanist' and 'feminine' do not support the usual fantasy of presence is surely to be cheered. Yet now that the phallus has jumped ship and, thus, indicates lack and not plenitude, the claim is both that the posthumanist feminine signals a common lack *and* that this move has not unwittingly fallen foul of the metaphysics of presence. It was Derrida's argument in 'Le Facteur de la Verité' that Lacan's conceptual framework of Woman, Truth and Speech precisely repeated the metaphysics of presence (the capitals are those of Lacan). A lesser-commented footnote in that long essay even remarked that this problem is further that of 'humanism' and that Lacan's related 'treatment of animality' is 'obviously of capital interest'.[42] We might also remember the structure of following remarked upon throughout *The Animal That Therefore I Am* that troubles the would-be presence of being – not I am, but I am following, not being-in-front-of but being-after. This troubling of human being, discussed in detail in the previous chapter, is foreshadowed by his earlier remark made during an exchange of letters with Verena Conley, on 'this unconscious that I am, that I follow'.[43] Such a remark recalls Derrida's overlap with, rather than overcoming of, psychoanalysis. Therefore, if I am (following) this unconscious, this *neither* displaces presence (into the reserve of an inaccessible Unconscious, *fort: da*, we might say) *nor* frames being as lacking.

Wolfe is at pains to reassure readers that this (Žižekian) version of a prosthetically supplemented real is one to which we humans are all subject, and, by virtue of its prosthetic rather than original, organic status, *we are all* posthumanist; *we are all* feminine. That is to say, Wolfe's theorization of this notion and this film is even-handed in its dispensation of lack. Though it might sound less unfair, it is the wrong solution to the right problem. A linked problem is the conceit wrought by Žižek that not only positions the phallus as lack-supplement but also determines the voice as the seat of 'radical ambiguity' and as something to which Derrida, appearances notwithstanding, 'remains blind'.[44] Žižek, thus, polemically positions himself with those who misread Derrida's critique of phonocentrism as a description of what voice is, period, that is, the living presence of the word in the instant of its delivery, echoing Erlmann's short-circuiting of the machine-event of 'Tympan'.[45] Crucially Derrida also solicits 'writing in the voice, the voice as differential vibration, that is, as trace'.[46] Writing here is, of course, in that displaced and transfigured sense of spacing. Žižek's conceit, however, in playing to what he calls 'radical ambiguity' then takes the voice in psychoanalysis as the stain of the Real, blotting the view that mistakes the subject for transparent self-presence.[47] Rather than the seal of auto-affective presence (the fiction of the voice in metaphysical thought according to Derrida), this stain of a voice is 'the greatest hindrance to the self-transparency of *Logos* ... in its *inert presence*'.[48] This 'hindrance' might form the allure of a posthumanist feminine through the vocal challenge of Björk as cast in *Dancer in the Dark*, but its inertia, its presence, as well as its preponderantly negative affect, should give us pause. Inertia brooks no spacing. Voice as vibration, trace, spacing, is on the alert for 'violent assignations'. Assignations, says Derrida,

> rings in its tongue in its idiom, very close to *assignation à résidence* [house arrest], to a judicial assignation, or ... to the regime of signification as if 'to assign' meant to assign to a subject, to the *law of the sign*, of the signifier or the signified.[49]

Spacing strikes the 'rebellious force of affirmation', resists 'sexual Difference with a capital D', resists being 'drag[ged] into a dialectic' and offers instead 'sexual differences in the plural' in 'open series'.[50] As such they are 'rebellious to any opposition'.[51]

While Wolfe provides an overall account of the complexities of *Dancer in the Dark*, his attention to a particular scene is given over to what is a relatively happy moment (prior to the murder and the inexorable path to execution), even as – perhaps because – the song that Selma sings refuses the suit of the well-meaning Jeff on grounds that she has 'seen all [she] needs to see'. With a principled understanding that the complex articulation of this film does not straightforwardly deliver a linear narrative and thus it cannot be judged wholly upon its ending, Wolfe sounds a note of resistance in this scene though reading Björk's open-mouthed and yet full-tongued vocal style as one that blocks the site at which her subjectivity 'should' be wrought (the mouth) – as one dependent on a specific relation to a man (marriage).[52] Opening her mouth to sing, her tongue blocks our visual access to her body in penetration by visual proxy. Wolfe thereby supports feminist independence, if we are persuaded by his reading of Björk's tongue as cinematically aberrant, given her widely recognized – and applauded – eccentric style, and if blocking language *qua* signifier is enough to redeem this film.[53] I will come to an alternative figure in the next section.

With Wolfe I agree that this film produces astute comment regarding cinema as neither limited by the visual nor engaged in any simple evidential showing of the visual. Casting Björk in the lead role clearly draws attention to voice in a way that a classic musical star such as Julie Andrews would not. Yet what is striking about *Dancer in the Dark*, in my reading, is the persistent call to rhythm. Each time Selma drifts away, she has used what seem to be insignificant or incidental noises to channel her daydream. After Selma has murdered Bill, she puts a record on the gramophone. However, rather than listen to music diegetically, it is the fluff on the needle that spurs another song (emphatically belonging to Selma's point of audition). In the factory assembly line where she works, the noise of the factory is re-orchestrated, loses function and becomes poetic. The machine handling sheet metal – bearing a striking resemblance to a printing press or tympan – in uncomfortable proximity to the woman who can barely see, may be a source of anxiety for the viewer. Selma is perennially close to getting her arm caught in it, yet on coming back out of her dream, it is she that has broken the machine and halted production (having inserted more than one sheet of metal at once).

Figure 6.1 Factory (*Dancer in the Dark*, 2000, Dir. Lars von Trier, Denmark)

While it does not spur one of Selma's 'own' musical moments, it is also worth indicating one of two scenes literally featuring a 'dancer in the dark': the scenes at the movies, when Selma and her friend Cathy (Catherine Deneuve) watch a musical together, Cathy taps out the choreography of the film with her fingers on the palm of her visually impaired friend's hand.[54] This touching scene is perhaps the closest that the film gets to an erotic encounter, and it underlines sound as the experience of hetero-affection, of being touched by the other.

Bar association

Given the vexatious relationship between symbolic and juridical law (between differently freighted relations between the contingent and the universal), I will now summon up the courtroom scene. Wolfe notes this scene largely for the 'unexpected and not entirely successful cameo' of Joel Grey as Oldritch Novey, the man Selma has persistently claimed to be her father, still living in Czechoslovakia.[55] Grey regularly turns up in films and television series in idiosyncratic cameos but is perhaps best known for his role as the Master of Ceremonies, convening events in the film *Cabaret* (dir. Bob Fosse, USA, 1972). In *Cabaret* the performances in the Kit Kat Club point up the incipient fascism of Nazi Germany.[56] Grey already carries, then, both musical and satirical extra-diegetic associations. Legal historian Pierre Legendre gives particular attention to the courtroom as a special case of psychoanalysis. Beyond the small fry of the private session, *this* is the clinical environment for healing rifts in the Symbolic order. Such a synopsis immediately indicates the Lacanian investment of Legendre's own theoretical framework. I refer to his work in light of that of lawyer and scholar of legal media Cornelia Vismann, whose attention to discourse evidenced her preference for Michel Foucault and attention to the citationality of media evokes that other 'Doctor of Law' – or J. D., Jacques Derrida.

With extreme economy, Selma answers the prosecution about her actions; she does not deny the murder and simply gives the explanation that Bill had asked her to kill him (as, in shame, he had done). Without fleshing anything out, it does not sound plausible and is easily used in the case against her. Her compact with Bill in keeping secret his theft of her money – 'I promised not to say' – has no persuasive force in a courtroom, the very place in which telling the truth (the whole truth and nothing but the truth) before the law should over-rule any other commitment. *Dancer in the Dark* does not bother to illuminate witness testimony with flashback. It allows the prior 'realist' sections of the film to stand; in so doing, it repeats the film's habitual complicity with evidence. Unusually, the cinematic audience to this film not only understand events according to the realist narrative but also witness what other characters do not, namely the musical sequences (thus, while the musical sequences are marked as proper to Selma, they also appeal to 'us'). Though we are later to understand that Selma's state defence lawyer does a poor job in defending her, *Dancer* does not show the case for the defence. Given that in due process this case should follow the prosecution, we, audience as jury, can only read Selma's song and dance *as* that case. In so doing *Dancer* departs from realist conventions, departs from the oppositional trap of prosecution-defence and performs an oblique address. I will return to this special defence.

From the prosecution's point of view, the only secret that Selma withholds is the identity of her father, the man to whom she has allegedly sent her every spare dollar at the expense of spending any money on her son (as a 'normal' mother would do, etc., etc.). Construed as an absence in the field of legal and symbolic representation, this secret is rectified by the court. Exposing this one lie makes it seem as if all she says are lies (a ruse explicitly used by the prosecution). Nor is this lie minor in the context of the coming together of state and symbolic law. Selma has invented the identity of her father and murdered a man who seems to all except Selma (and the cinematic audience) to be an exemplary father, rendering her deed close to that given particular, indeed exemplary, treatment by Legendre: patricide. Moreover the prosecution is quick to imply an anti-American aspect to this crime, committed by an East European immigrant and probable communist who prefers her homeland. Vismann, elaborating upon Legendre's frame, suggests that the court's function is to re-institute the violated linguistic institution that suffered through patricidal affront. Effecting closure on a grand scale, '[t]here, in the legal frame, every thing and every person is at the right place.'[57] This underlines why the courtroom scene is not one among others but a significant one to this film. In the paradigmatic case of patricide that Legendre analyses, he further suggests that the accused himself desires his own restitution as subject in and through the confession to murder.[58] Selma, we should note, neither takes pleasure in nor places special emphasis upon the killing of her neighbour during her courtroom testimony in response to the prosecution.

To everyone's surprise, Novey – now a California resident – is summoned by the prosecution. He announces that he does not even know Selma, let alone be related to her. As the prosecution persists in drawing out Novey's evident non-relation to Selma, the film cuts to her craning her neck around to listen and resulting in a medium close-up in which her ear is centralized. Selma is not, however, listening to either the prosecutor's triumph or the real Novey speaking about his real circumstances, words that might be experienced as at the very least humiliating and at worst as shattering in their exposure of her fictional support. Rather, she is listening to a sound that does not count

Figure 6.2 Hear (*Dancer in the Dark*, 2000, Dir. Lars von Trier, Denmark)

within this space. The courtroom artists sit behind her (thus their activity is not visually significant in the architectural arrangement of the court): their pencils scratch against their paper as this scene is gradually doubled by drawings. Selma is listening to the sketching of the scene, to its stereography. This graphic inscription – now louder than any voice – generates the rhythm that taps into her fantasy and redraws the court.

We can now shift the attention that Wolfe finds unavoidable regarding Björk's tongue in the 'I've seen it all' number towards her ear in this one. With Derrida, 'Why don't we turn our ears toward a call which addresses and provokes above all else … To turn one's ears to the other when it speaks to "whom," to "what," to this "who" which has not yet been assigned an identity'.[59] Instead, then, of searching for a figure that would block passage to the vulnerable feminine, her ear appears as the channel by which Selma increases the volume of marginal noises. While psychoanalysis might perpetuate the notion that the ear is a 'hole', even one that is impossible to close or block, we have seen from Derrida's 'Tympan' that this is not exactly the case.[60] To 'forget' the complex play of the 'indefatigable' tympanizing ear would be, he suggests, to 'cry out for the end of organs, of others'.[61] Of course there is an overlap. Not between the psychoanalysis of the hole and the deconstruction of the spiral but of the vestibule of the ear and the vestibule of the vulva.[62] Not content to remain within human biology, the note that touches upon the vestibule opens an archive of thought:

> Tympanum, Dionysianism, labyrinth, Ariadne's thread. We are now traveling through (upright, walking, dancing), included and enveloped within it, never to emerge, the form of an ear constructed around a barrier, going round its inner walls, a city, therefore (labyrinth, semicircular canals – warning: the spiral walkways do not hold) circling around like a stairway winding around a lock, a dike (dam) stretched out toward the sea; *closed in on itself and open to the sea's path.*[63]

In Vismann's analysis of the history of media technologies within the court – from its theatrical mainstays to increasing televisualization – she notes that this is no simple extension or modernization but more frequently viewed as competition. The increasing pull towards media framed as essentially evidential rather than hermeneutic produces competition with the work of judgement and the 'process of memorising'.[64] Vismann's recognition of deconstruction, however, allows her to obviate the legal fear that the incursion of televisual media necessarily predicts a degeneration of authority and hence the possibility of justice. While Legendre holds that the 'replaying of the crime' in its media instantiation in the courtroom (through photographic or video evidence) becomes the prompt for the restitution of the subjectivity of the accused, Vismann implies that this 'replay' is neither bound to such media nor of guaranteed effect. Rather, there is a replay – or citation – of images already at work in the court as theatre that video techniques 'consummate' but do not invent.[65] In the case of *Dancer in the Dark*, it is the activity of drawing that places citation at the heart of the court. Instead of that drawing's principal function being the representation of what is in front of (or present to) the artist in the service of the judiciary, the scratching sounds that this activity also produces are magnified to usurp the expected order and initiate another rhythm. The pencils cease to point towards a representation: erasing the present, erasing the sense of virgin matter, of 'first' inscription, they conjure 'the weave of differences of forces'.

Re-describing the court and what might happen there from the moment attention settles on the drawing – then tapping – artists, everything changes. The heads of the jury and audience alike start nodding in unison, saying a different kind of yes, synchronizing a rhythm across the whole courtroom. This is the scene in which Selma wants to star. She stands up and starts singing her love-song to her purported father. Absconding from her allotted place as defendant, she explores the architecture of the environment, walking atop the bar delineating the space of the jury like a gymnast. Thus, she also walks over the whole legal profession metonymized by 'the bar', and walks over the division of sense from signification issued by the Lacanian 'bar'. One by one the members of the court leave their places and join in, and the view of the court as an organized theatre anchored by the figure of a judge gives way. Indeed the elevated bench that should serve only to heighten the authority of that judge as separate from and superior to the court that he oversees is transformed into a stage on which Novey – and then Novey and Selma together – tap dance. Tap-dancing: a musical show-time stalwart, delivering a series of blows as it taps out its own rhythm. Tap-dancing: that which was barred from the town's rendition of *The Sound of Music*. The rest of the people in the room support this new eventuality, clicking their fingers, clapping and slapping the furniture around them. Even the bemused judge finds his movements falling into this new state of affairs, and the now nodding district attorney (Zeljko Ivanek) is disoriented. Judgement is suspended and the court ceases to divide into advocates of prosecution and defence. This song and dance only stops when – as Selma's song repeats the fantasy that her father would always be there to catch her – the court announces the verdict of the jury and she is sentenced to death. Cutting her off, the guilty verdict predicts death by hanging, and no one being able to stop this fall.

This scene is unique in terms of both court room drama and a repetition – in *Dancer*, every time Selma tunes into her other world, a similar pattern takes place, one that unhooks those around her from their everyday laws, even disbands natural law (the murdered Bill rises to dance with Selma). Every time Selma rearranges her environment into one that is not bound to a certain end. It is not that her daydreams present a space of no time, nothing, pure absence. Rhythmically differentiated, they cannot but mark timing but without it inexorably leading to a certain organized end. Faced with a path of 107 steps leading precisely and notoriously to the place of execution, Selma makes a song from listing numbers out of sequence so that these steps – these '*pas*' – do not count, do not progress, will not come to the scheduled end. In the courtroom she produces a scene that is not exactly contempt of court, since there is no outright challenge or derision of the proceedings that would oppose the prosecution, derisory though it might be, but it is rather the cinematic *tympanizing* of it. 'Publically' ridiculing the court only in the sense of appealing to the cinematic audience, *Dancer in the Dark* decries the limits.

Writing of the tympanum, Derrida asks whether 'to transform what one decries (tympanises), must one still be heard and understood within it, henceforth subjecting oneself to the law of the inner hammer?'[66] This 'inner hammer' is one of a chain of small bones in the inner ear found on the internal surface of the tympanic membrane. It both transmits sounds and mediates them, softening those that might set up too energetic, too painful, a vibration.[67] Drumming on the limits of philosophy or indeed a court of law, one risks thus being filtered out. If we remained with Selma's narrative, there would be no reverberations from her dream of more rhythm: after breaking the machine at work, she is fired; after tap-dancing on the judge's bench, she is sentenced to death.[68]

Enunciation of the death penalty

I turn now from the interchange of prosecution and defence in the court of *Dancer in the Dark* to the last sequence: the execution of the sentence. This is the scene with the most pressing ethical stakes in which there is greatest overlap between the worlds of this film. Careful attention to the texture of *Dancer* shows the film struggling with the repercussions of the philosophical tympanum on which Derrida insists.

The execution of Selma – this laughable case of a 'non-criminal putting to death' – relentlessly pursues the vicious formalism of law's enactment.[69] That is, the rote repetition of law negates the possibility of justice. The transposable figure of the 'non-criminal putting to death' as it habitually applied to those we call 'animal' is discussed at length in the first and last chapters of this book. Here its force captures the one called both 'criminal' and 'immigrant' before the law of the United States in order to bypass the ostensible transgression of the primary injunction, 'Thou shalt not kill'. While they are not wholly directed at the United States, Michael Naas has suggested that while Derrida's monumental 1974 volume on the family, law, sexuality and blood as they are given in metaphysics, *Glas*, implicated the France still perpetuating the death penalty (by guillotine), his more clearly abolitionist seminars on the death penalty conducted twenty-five years later confronted the United States.[70] Amongst developing the poetic grounds for resistance to the death penalty in the absence of a philosophical one, those seminars frequently point out its disproportionate application to poor, black citizens, especially black men.[71]

Judged, sentenced with the most enforceable performative, Selma is to be hanged until dead. But she cannot wait for her death in the manner prescribed and collapses at the scaffold. A female guard, Brenda, who has already gone beyond a merely functional relation to Selma through sympathetically identifying with her status as a single parent, gives assistance that persists in treating her individually, and not simply as the one – like any other one – found guilty before the law. It should be noted that another female guard shows no such sympathy, disbanding any notion of an essential feminine good. Part of the enactment of state execution by hanging, as shown in *Dancer in the Dark*, requires the condemned to stand, demonstrating life before death, full consciousness before death. As Derrida writes, 'in the figurative sense ... execution always attacks the head ... everywhere a prohibition is observed against executing a sentence on someone who is out of his head, as they say ... He must die awake'.[72] In the face of Selma's collapse, officials calmly, even impatiently, fetch 'the board' by which prosthetic means her now irregular body is straightened out and rendered erect (no margin for error in execution). Again, through the vile farce of disavowing the spectacular quality of the execution, the state attempts to hood Selma, thereby effacing her face, making her dead already in the uniform of the executed. But as she continues to weep and complain in panic of not being able to breath, Brenda – also now in tears – removes the hood, countermanding those colleagues who adhere to due procedure. In the brief interlude of this irregularity, a senior official appeals for legal confirmation by means of a telephone call: can the blind woman remain unhooded for her own execution? The telephone, in the theatrical time of the film as well as in US penal institutions, opens a space for a possible pardon, given that it connects 'the place of execution to the mouth and the ear of the sovereign governor, keeping it in tele-technic relation with the transcendent place of sovereignty, with the governor who holds the quasi-divine power of pardoning'.[73] While this one will authorize Selma's execution, authorize letting the 'line' fall, it opens a space for Cathy to also abandon her place as witness, pushing past security to tell Selma that her son, Gene, has had his operation and now can see. Cathy gives Selma the one thing she can at this moment, which is to tell her that she was right to sacrifice herself for the sake of her son (whether or not Cathy believes this herself). If there is any progressive relation between justice and the feminine in *Dancer in the Dark*, it comes through the actions of Brenda and Cathy. Brenda, moreover, is the last person remaining in the execution room (Cathy having been forcibly removed from proceedings).

Having previously uttered nothing to the formal permission for any last words, Selma, 'standing' with the noose around her neck, 'actually' sings. Holding the unexpected talisman of Gene's now-redundant glasses, to the mortification of all, the one who should have already been silenced now sings. Right next to her death, Selma sings a song of deferral. Tempting as it might be to counter-pose her living voice to this imminent death, the labour of her breath, her magnified heartbeat and then her soft whistle, from which rhythm she generates her song, should already meter that voice as one of 'life death'. Moreover, this is her favourite genre of song: the next-to-last one of a musical.[74]

Selma is always next to the last (to death) since rhythm is an invitation to spacing and she has requested 'more rhythm' from the start. While some shots take in the officials, Brenda's empathy, others in agitation at their lack of instruction, Selma's eyes are closed and the film sympathetically implies her world through extreme close-ups that largely exclude the noose. This does not mean that no one hears her singing: the whistle catches Brenda's ear and she turns to listen. As with the abrupt break with Selma's 'defence', the law again slams into Selma's deferral: the mechanism of the scaffold violently and loudly breaks with her song as it breaks her neck and the film's soundtrack. In a quick succession of shots, *Dancer* even shows the hung woman on the suspended board followed by the glasses now broken on the floor, needed neither by Gene (the heir who now can see) nor Selma (who absolutely cannot). As this 'letter' arrives at its destination in an overkill of the stigmata of castration, can we retrieve a posthumanist feminine from this act of suicidal refusal to help her own case by using her saved money to hire a competent lawyer instead of on her son's sight?[75] To do so would overly invest in Selma's own actions (as maternal self-sacrifice) and step back from the broader texture of this film – the 'site' for any ethical redemption of this film.

Cutting Selma off in this brutal manner is not quite the end of this film, which continues, silently, in a complex interplay of formal strategies. On the one hand, Selma's body, rendered phallic by the board, falls abruptly through the trapdoor, cutting through the floor downwards to the viewing room below (the audience only 'directly' share space with the subject of the death sentence post execution). This space is revealed as her final stage: the obscene theatre of the jail replete with curtains. Again, given the trapdoor (as hole) and the unifying restraint of her body by the board (as phallic), the obscene stigmata of castration dress this stage, exchanging a life for a life: death penalty as social contract.[76] Lest we then take that 'stage' to be Selma's desired end for her story to be complete in this event in the 'dark' in which she can 'dance' forever, the writing of her song continues as a typographic inscription across the screen. Transcribed into text that effectively crosses out her dead body, *Dancer* cites Selma thus:

> **They say it's the last song**
> **They don't know us, you see**
> **It's only the last song**
> **If we let it be.**

Not the neutral officialdom of Times Roman or the typewriter reportage of Courier, *Dancer* uses a slightly kitsch font – a marked text in speech marks affected by a style.[77] Crossing out not the body *tout court* but Selma's phallicized corpse in particular, *Dancer in the Dark* countersigns the ambiguity of which she sings joining that luxation to an ethical appeal, even as the jail bureaucrats verify her death. The last song, the text implores, need not be so. 'We', the cinematic audience doubling as an alternative jury, are implicated in the decision as the camera aligns us with the viewpoint of the witnesses to this event only to track back upwards, after the curtains close, to the execution chamber in which only Brenda remains. Standing in a place that no one should legally occupy, in the last diegetic frame, Brenda's head is bowed in mourning for *this* woman, for Selma. The camera continues to evenly track upwards through the ceiling to the black frames of the credits. Sound only recommences – with Björk's track 'New World' – once we leave the diegetic scene and the credits roll. Reminiscent of the individual images on a strip of celluloid divided by the floors, the film announces itself *qua* film distancing itself from the realism – and the judgement – of the legal theatre just witnessed. 'New World' begins: 'Train-whistles, a sweet Clementine/Blueberries, dancers in line/Cobwebs, a bakery sign… '. B/eating the limits indeed.

Playing it by ear

Invoking the structure of the manual printing press, kin to the mystic writing pad, Derrida asks after its multiplicity of tympanums, 'which is to upset the entire space of the proper body in the unlimited enmeshing of machines-of-machines'.[78] This is not a techno-fetishism. In the last sequence of *Dancer in the Dark* film and theatre (the latter in both dramatic and legal and penal contexts), voice and typography make their impressions. All types of text matter in this remarkable film, and, all texts *type*: they strike a blow. Types, however, vary. This does not mean that there is simply an increase in the number of discrete types now due their own place replete with legal representation. Neither is the technicity suggested by all these varieties of impression a belated addition marking the special category of the human in which the supplement is ring-fenced to signification. The sensible world cannot sustain the fiction of autonomous virile masculinity as subject, including in its inverse form as universally castrated signifier for another, humanized, signifier, before the law. The typing texture of the living proposed in deconstruction cannot be immunized from non-human bodies. As Derrida remarked in 'And Say the Animal Responded',

> the structure of the trace presupposes that *to trace* [another impressive gesture] amounts to *erasing a trace* as much as to imprinting it, all sorts of sometimes ritual animal practices, for example, in burial and mourning, associate the experience of the trace with that of the erasure of the trace.[79]

In this deconstructive gesture, any appeal to a 'posthumanist feminine' can suffer no relation to lack. Derrida himself 'rarely speak[s] of lack'. He writes that lack 'belong[s] to the code of negativity, which is not mine, which I would prefer not to be mine. I don't believe desire has an essential relation to lack'.[80] Dialectical synthesis can take an oblique turn, including that maintained in post-Lacanian psychoanalysis. The posthumanist feminine suffers only in the transfigured sense of constitutive vulnerability, the 'ability to suffer' discussed in the previous chapter. As expressed in this one, in this film, such an event requires a strong reading to draw *Dancer* away from the cruelty to which it plays. If this chapter dwells on the deferral of the next to last song, the next chapter is lieu of conclusion.

7

In lieu of conclusion: The cardio-pedagogy of *White God*

I want to cut to the chase. I want to talk about endings. How will it all pan out? The most arresting aspect of Kornél Mundruczó's extraordinary film, *White God*, is the last scene, the ending that is not one.[1] The end of this chapter, and of this book, is not to 'spoil the ending' of this film in particular but to amplify what I understand as a striking dismantling in lieu of conclusion. In the very place of conclusion: the slaughterhouse.

White God's cinematic trailer acts as a synopsis.[2] Like the film's poster it almost gives the ending away but holds back in a manner that crucially, if misleadingly, maintains the positions of the young girl as elevated human and her dog as an exceptional pet (although the poster is one step more familiar since the hooded girl it features registers only as a 'child', her sex going without emphasis and thus liable to be read as masculine). The sheer thrill and the energy of hundreds of dogs running through urban streets may at first recall Swedish House Mafia's pop video *Save the World*.[3] However, where those show-ready pedigree dogs 'save' or redeem the world from bad men, simply acting as alibis for ideal humans enforcing the good in their every encounter, *White God* poses a more complex investigation into what Derrida calls the 'force of law' and the possibility of justice. In brief, while the two terms might be thought to be interchangeable, or that law should concern itself with justice alone, Derrida points to both their separation – as the rote application of legal precedent versus the singular invocation of justice – and their interrelation, not least through what he deliberately calls the 'ordeal' of the decision:[4] 'A decision that would not go through this ordeal of the undecidable would not be a free decision: it would only be the programmable application or the continuous unfolding of a calculable process. It might perhaps be legal; it would not be just.'⁵ This 'ordeal' is named deliberately to counter impoverished readings of deconstruction as the 'free play' of the signifier or reification of undecidability as a permanent condition. It also exposes deconstruction to the heart of justice. Allegorizing forces are hardly absent from *White God* – its very name makes that clear – and the film is far from naïve regarding its possible interpretations. But where allegorical or metaphorical procedure habitually says 'this *is* like that' or 'this is *really* that' disappearing a comparative term (animals) in favour of the anchoring one (humans), *White God* convenes all the terms and exposes their hierarchical operations. Moreover, as this chapter will elaborate, *White God* proffers an astonishingly precise challenge to the virile subject of carno-phallogocentrism mapped into the structure and the address of the *mise en scène* at the end.

Endnotes

In classic cinema the function of narrative closure is to restore equilibrium, to restore the order of things, resolving the disturbance initiating that narrative. Historically, the work of closure has not merely solved a problem but has done so with an ideological finesse. The 'woman-question' was regularly answered by classic cinema through the restoration of normative heterosexuality including the economic dependence of the female protagonist upon the leading man. In her well-known diagnosis of the Hollywood cinema of the 1930s and 1940s, Mary Beth Haralovich observed, 'If a woman is in a non-normative role in economic control and production, she will cede that control to a man by the end of the film.'⁶ Feminist film scholar Annette Kuhn remarked that not only was this recuperation of normative heterosexuality habitually embedded in cinematic closure but that it was a strong structuring element of mainstream narratives throughout. Worse, and as witnessed in *Dancer in the Dark* discussed in the previous chapter, any woman failing such recuperation was liable to be 'punished for her narrative and social transgression by exclusion, outlawing or even death'.⁷

The story of *White God* opens with fragile domestic relations: a divorced family sees the woman leave for Australia in the company of her new partner.

He is a professor, financially and intellectually buoyant. We are briefly introduced to their world as romantic, sunny and bucolically figured around the woman's daughter playing with the family pet. Their separation could not be more violently pronounced than with what can be called the film's establishing cut to the carcass of a cow being sliced open in an abattoir: the cut of the edit underscored with a corporeal cut moving from the guts to the heart. The family: the split. The animal: the cut. The heart. The brutal sequencing of these scenes makes sense when we realize that the woman is leaving her daughter, Lili (Zsófia Psotta), and the dog, Hagen (Luke and Bodie), with her ex-husband: leaving them, jarringly, at the place of work that it is his 'house', the slaughterhouse. The father (Sándor Zsótér), a former professor, remains in a skilled profession, even if he shows no pride or enthusiasm for his new position as an inspector of what is good to eat, even if his heart isn't in it. Crucially, the film identifies him not with the blunt force of the act of slaughtering cattle but with the judgement of what is 'fit for human consumption', countersigning the life-or-death distinction between human and animal with every decision. He is thus made integral to both the practical and the symbolic maintenance of this industry along with what remains of the family: his 'ex's' last words to Lili are the instruction that she should 'obey' her father. Thoughts that the narrative arc of this film will be to reunite the human family are dispelled when the father casts Hagen out onto the Budapest streets. In the midst of a crackdown on 'mixed-breed' as opposed to properly 'Hungarian' dogs, this eugenic directive means that Hagen loses the legal protection of fictional purity and faces being rounded up by dogcatchers and euthanized. This horror, however, pales, next to the intimate instrumentalization of Hagen by the dog-fighting ring for which he is so purposefully selected, because he 'still has a heart'. *White God* repeatedly issues threats to the heart. This is a film that, at heart, acts on the heart, which speaks of and speaks to the heart even as it stands before the law.

One reviewer has located *White God* within the niche genre of films featuring animals 'journeying home' (such as *Lassie* or *The Incredible Journey*), and Erica Fudge has suggested that it is the function of the 'pet' to 'come home' (indeed the trailer implies this type of reunion, albeit with the seductive novelty that this anthropocentric and proprietorial maintenance of borders is affirmed by a young girl).[8] But this place is unmistakable: as daughter and dog are dropped off, the quotidian pattern of workers carrying exsanguinated bovine

Figure 7.1 Courtyard (*White God*, 2014, Dir. Kornél Mundruczó, Hungary)

carcasses passes in the other direction. Coming home to the slaughterhouse obliges the animal question, revolving around the status of companion as much as livestock animals and the human family's performative participation within such divisions. *White God* is closer to horror genres, such as what Gregersdotter, Hoglund and Hallen name 'animal horror cinema', and the fear of predatory inversion typical of such films.[9] That is to say, when animals revolt against humans, horror is frequently figured through the reversal of what we take to be civilization: instead of eating animals, humans will be eaten by them.

In the broadest strokes, *White God* does commence a canine revolution in which all the dogs rise up against all the humans.[10] It is like a cross between *Spartacus* and *Rise of the Planet of the Apes*.[11] It knowingly marshals the spirit of resistance, and the righteousness of such resistance. But the dogs do not predate humans as food. Their revenge strikes back, yet they do not take humans for meat. While, as Dinesh Wadiwel argues in *The War Against Animals*, 'The terror of the rogue animal surely is that in resisting human violence, in levying a violence that might be interpreted itself as instrumental, this rogue animal would be sovereign',[12] the rogue state of *White God* points beyond a simple exchange of figure at the apex and thus intervenes in the substitutability of beast and sovereign.[13] In so doing the film intervenes in the normative workings of allegory as they inform the political state and ethical relations. Beyond reunion, the film is also finally beyond retaliation, even as the narrative pull off much of the film leans on the calculation of punishment as debt that is *jus talionis*, or talionic justice.[14] Yet the film underscores, returns to the very space in which bodies become meat, in which the heart must be inspected – the courtyard of an abattoir still within the city streets, not yet displaced to the anonymous giant sheds far away from urban centres that characterize the animal industrial complex in 'advanced' capitalism. That it does so in tandem with an uncertain family unit with a diminished paternal figure gives us pause for thought. Where is the exit? Can one exit the logic of the slaughterhouse? Can one put an end to the end? Can one exit the end and survive? These questions can be heard both rhetorically in the habitual sense that would absolutely distinguish between the outcomes for humans versus non-human animals but also, crucially, in more far-reaching ethical terms, given the film's endeavours. The heart of this chapter opens towards these questions. To meet them, it will give privileged attention to the critical orchestration of the *mise en scène* of the end and the heart-stopping gesture that calls a halt to the narrative that *White God* has led viewers to expect. In so doing the expectations of genre glossed here also drop away for a more surprising engagement with poetic address. If that seems like a curious *non sequitur*, consider the fundamental movement of force and signification across Derrida's work affecting not just the force of law but also the question of 'what is poetry'.[15] In his essay of that name, there is a 'heart down there, between paths and autostradas, outside of your presence, close to the earth, low down'.[16]

Curled up in that vulnerable event 'close to the earth' that Derrida calls the poematic is a form of address that desires to learn by heart in which both rote repetition and singular cherishment come together. If *White God* may be said to be pedagogical, it is not in a didactic sense but one that inspires learning by heart. Ahead of this attention to the end and the manner of its demise, I want to consider what's in a name.

(White) God/Dog

Above or beyond the law, alternatively in violation of it, the political figures of the beast and the sovereign seem to form two distinct poles while sharing what Derrida calls a 'troubling resemblance', this 'being-outside-the-law', this 'reciprocal haunting'.[17] Indeed this uncanny proximity cannot but be heard in the French title of Derrida's seminars, *La bête et le souverain*, where the difference between *est* and *et* ('is' and 'as') is inaudible. *White God/White Dog*.

Legal theorist Colin Dayan has written a lengthy study of the interwoven spectral and material, animal and human subjects before the law: her book is called *The Law Is a White Dog: How Legal Rituals Make and Unmake Persons*.[18] Canine figures that do and do not involve actual dogs arise in various contexts across the book without congealing into a single sense. These figures vary from military dogs, spiritual guides, legends involving cures from rabid behaviour, Haitian ghost dogs in which their 'white skin' indicated no skin at all, but rather that a 'person's spirit remains immured in this coarse envelope, locked in the form of a dog', and as a name for the metonymic identification of dogs trained to hunt slaves in antebellum America with the hooded appearance of members of the Klan.[19] In the latter context, Samuel Fuller's 1982 film, also named *White Dog*, cannot but be brought to mind.[20] This was not a deliberate point of reference for Mundruczó, but a film that he was pleased to belatedly align with in solidarity against racism.[21] *White Dog/White God*.

Both films address justice, both involve forms of hesitation, one of them re-consolidates a known horizon, while the other does not know, cannot tell, where it is going: one of them exposes the heaviest weight of human projection upon the flesh of others, while the other makes another offering altogether. The greater solicitation of the law – in the Derridean sense or insistence upon an etymological '*shaking* in a way related to the *whole*' and so the 'shaking' of the law[22] – is the one that recognizes that

> If we wish to speak of injustice, of violence or a lack of respect toward what we still so confusedly call the animal – the question is more current than ever ... one must [*il faut*] reconsider in its totality the metaphysico-anthropocentric axiomatic that dominates, in the West, the thought of the just and the unjust.[23]

Fuller's *White Dog* was not officially released to US audiences until 2008, after rumours at the time of its production suggested that this was a film that was itself racist – rather than critically about and indeed against racism.[24] However, the narrative revelation that the film's rescued white Alsatian does not turn out to be simply a trained 'attack dog' but more precisely and insidiously a 'white dog' trained to specifically attack black people moves markedly away from the story in the novel on which the film was based (*Chien Blanc*).[25] Where the novel had this dog retrained by a black Islamic fundamentalist, not to be cured of his acquired racist aggression but to invert it in order to now only attack white people, the film creates a much more sympathetic black trainer who truly attempts to train racist violence out of this dog, and by extension out of the world. The way in which this retraining fails is an index of *White Dog*'s ambiguity regarding what justice might entail were it to treat the question of race and the question of the animal as intimately linked problems. Essentially the concluding scene and 'final test' involves the trainer, Keys (Paul Winfield), inviting the dog into an arena-like cage but this time without any protective clothing (the dog remains un-named, other than with occasional literary nicknames such as 'Mr. Hyde').[26] While the dog runs towards but does not attack either Keys or Julie (Kristy McNichol), the young white woman who rescued him, he then turns on Carruthers (Burt Ives), the elderly white man who runs the remote 'Noah's Ark' animal training environment, in which they are ensconced. That he attacks Carruthers and must be shot by Keys seems to mean that the effort to 'unlock' this dog has proved too much; this is simply an unstable and dangerous dog who, by right, must be 'put down'; Keys is a good man who will execute the law by preventing this dog from killing any more human beings, having earlier killed two black men. The law in the form of local police was never called upon regarding these prior fatal attacks. Rather, there was an extraordinary scene during which Keys, Carruthers and Julie colluded over dinner in favour of their continuation of the secret retraining

Figure 7.2 Toast (*White Dog*, 1982, Dir. Sam Fuller, USA)

exercise without involving the police or animal control officials. Acting extralegally in this space hidden away from the ordinary world, they commit to the prospect of eliminating racism by technical means. Yet they maintain the law as a carnivorous one: in a room on the ranch chock-full with animal skulls – it is practically a Natural History Museum – they yet toast 'to the hamburger' (with this toast closing this scene).

Two elements – one of sequencing and one of framing – introduce a reading other than the 'failure' of the final test, albeit one that is not fully articulated: the immediately preceding scene reveals the original trainer who made this dog into a 'white dog'. He is an elderly white man, derided as the racist he is by Julie, who demands that he does not instil the same 'sick' violence in the 'puppies' with him. The 'puppies' are his two daughters. Conceivably, then, the dog recognizes his original trainer in Carruthers and is attacking this training, not a random person, not white people in general. Conceivably the dog is exacting a form of *jus talionis*, issuing the death penalty to the memory of the one who had taught him to illegally kill.[27] The second element arises in the cinematic compassion afforded to this dog. Instead of moving away into the distance 'with' the three protagonists (Keys, Julie and Carruthers) united in a future horizon in and of the human, with Keys as the one who is redeemed by virtue of upholding the law and shooting this dog, the last shot remains mournfully 'within' the arena close to the body of this now dead dog, the sacrificial beast for an imaginary white sovereign.[28]

Both *White Dog* and *White God* feature forms of animal training. While *White Dog* attempts to rectify the racist work of the bad trainer with a good one – a bad Father with a good one – in preservation of due human sovereignty, *White God*'s underground traffic in dogfighting operated by gypsies inadvertently produces an army, *pace Spartacus*, to fight under something like the name of the animal. Yet the spectral contamination of Dog and God, beast and sovereign bars this fight from really changing the order of things. Where Derrida's affirmation of the practice of deconstruction in the closing pages of 'Signature Event Context', as that which 'must, by means of a double gesture, a double science, a double writing, practice an *overturning* of the classical opposition *and* a general *displacement* of the system', the reversal of beast and sovereign stumbles over this crucial second clause.[29] At an angle to this reversal of terms is Lili and her promise, made early in the film, in observation of a man bluntly instructing another dog to 'sit', that she would never do such a thing.

It is a truism of Freudian and Lacanian psychoanalysis that culture is founded upon a crime: the primal murder of the father yields not only the beginning of law but the beginning of criminality as such through the restoration of the forbidding Father as a symbolic figure. Feminist readers of *Totem and Taboo* might readily alert us to Freud's vacillations on the maternal and question just how these formidable beginnings as paternal, as properly patriarchal, were entrenched (as discussed in the first chapter of this book). They might also allow us to trace Freud's hesitations as to the being of this founding figure, the patriarch, and thus the question as to whether our origins are human or animal. The drive of *Totem and Taboo* is, of course, to end squarely on the side of the former. The logical corollary here is that if law and crime arise as foundational to human culture up to and including the death penalty as paradigmatic, then animals do not commit criminal offences. Not only this, but in Jacques Lacan's formulation of this argument, when humans commit violent offences against other living beings, their entire enterprise operates at the level of the signifier. Thus they only actually offend against other humans. In *The Beast and the Sovereign*, Derrida draws our attention to the pernicious work of the 'fellow' or '*semblable*' in Lacan. The latter writes, 'This very cruelty implies humanity. It is directed at a fellow [*un semblable*], even in a being of another species.'[30] Lacan's formula gives rise to the most aggressive allegorical conscription of the bodies and the figures of other species. A 'what' may be the ostensible means of cruelty, but a 'who' becomes the signified target.[31] Animal blood services human cruelty. There is thus no 'crime against animality' whether amongst animals or towards animals.[32] As Derrida laments, this fraternalism of the 'fellow'

> frees us from all ethical obligation, all duty not to be criminal and cruel, precisely, with respect to any living being that is not my fellow or is not recognized as my fellow, because it is other and other than man.[33]

From the moment that Hagen breaks free from the men who had stolen him and brutalized his body and spirit into one to be abused in dogfighting, a moment repeated when the effectively radicalized Hagen subsequently leads a break out from the dog 'shelter', *jus talionis* appears to be endorsed by *White*

God. Writing of the justification of the death penalty in the thought of one of those central to the Enlightenment, Immanuel Kant, Derrida remarks that for him, *jus talionis* should not be

> in principle, in law, a horrible vengeance, but the reference to an impersonal principle of reparative justice that, precisely, does not obey the subjective and egotistical and impassioned or impulse-driven interest of vengeance. No more than a tooth for a tooth, no more than an eye for an eye: this is the beginning of justice or right in talionic law.[34]

While this juridical exchange of supposedly equal matters should not exceed its economic mandate but, rather, should be 'detached from any interest', in *White God* the punishment exacted by the dogs risks enjoyment.[35] Their actions do not exceed the apparently simple exchange of a life for a life, but the enjoyment marshalled by the film exceeds punishment (assuming that crime and punishment can be adequately measured against each other – certainly for Derrida they cannot). Dogs who were forced to fight each other to the death; dogs who were judged to be impure, unclean, deserving only of death; dogs who were condemned to the so-called euthanasia now pound ecstatically through Budapest. That 200 dogs flood the emptied streets – producing a singular image of the city under canine occupation – is in itself intoxicating to watch.[36] The driving bass on the film's fast-paced extra-diegetic score matches the pace of the dogs and their excitement, inciting our own speeding heart rate as it brings the dogs and the audience into alignment. As the film's advertising line has it, drawing a figurative alliance with and strength from all those cast as 'underdogs', 'The Unwanted Will Have Their Day'. Moreover, the opening scene of the film promises the penultimate ones of hundreds of dogs running through Budapest, with Lili, on her bicycle, ambiguously leading them or being hunted by them. A news report within the film tells us these rogue animals are not behaving 'like dogs' but are acting in concert. All roads lead to retribution in their charge: we see them seek out and kill the man who sold Hagen into the underground traffic in dogfighting; we see them kill the man who 'trained' Hagen to fight; we see them track down the landlady who first told Lili and her father that 'mixed-breed' mutts were not welcome and blood on the floor later evidences her death; we see the dead body of the butcher preceded by a shot of a horizontal slab of meat looking uncannily like a human head and torso. Within their evident itinerary of 'pay back' the dogs' restraint at the butcher's – in that they do not take any meat – seems to gesture towards a wider interspecies solidarity even before the showdown at the slaughterhouse: these dogs do not systematize the bodies of other animals in order to eat meat. On the other hand, it ties in with the purported moral exactitude of *jus talionis*: the dogs kill the killer, but they do not take advantage of this punishment by becoming thieves. Lili – on finding the bodies of the landlady and the butcher – recognizes that their purpose is not random and that consequently her father is in danger at the slaughterhouse. While his individual 'guilt' lies in having abandoned Hagen to the street (and thus likely to his death with or without his conscription into the dogfighting ring), the father's metonymical identification with the slaughterhouse is already established through his work and the dot of blood marking his shirt when his character is first introduced.

A terrible performative power of the 'cannot' fights against the film's narrative just recounted: 'But animals can't.' This ironically reactive Cartesian argument is corrosively poised to activate when animals, whether in the world or in the deliberations of a film such as this, step beyond their allotted containment within the field of those who do not have the ability to respond, whether taking the form of to speak, to lie, to laugh, to point, to give, to grieve, to dress, to die. As detailed earlier in this book, it was the major work of *The Animal That Therefore I Am*, not just to trouble Descartes' assurance of an absolute distinction between reaction and response mirroring an absolute distinction between the animal and the human, not just to release the animal from the containment of the force field of the 'cannot', but to question our own assumed possession of ability. While Derrida moderated this professed ability into the vulnerability or weak power of being able to suffer in that book, in 'Force of Law' we might align this move with the 'experience of the impossible' that is justice. Gil Anidjar names this precisely as a 'force of weakness'.[37] Again this is a counterintuitive move, but one that is necessary if there is to be any justice worthy of the name.

But animals can't. In this context, dogs can't act in concert, can't act to plan and – most of all – can't recognize death and thus can't recognize a crime for which they might execute just punishment in the form of the death penalty. While the film shows them doing exactly these things, such is the mono-directional force of allegory as anthropomorphism that the narrative is liable to be read as really about humans anyway. To claim or dismiss the film as a straightforward allegorical duplication of humans by dogs accepts the figure of the 'underdog' as one that is never about dogs but rather always in reference to benighted humans treated inappropriately, treated *like* dogs. To overturn this species of mistreatment then is to become jubilant (as Lacan might say recalling the infant's seizure of the mirror image[38]), in distinction to a more negative reading of our affirmation of the canine uprising as a suicidal identification, given that 'we' are faced with an army of dogs who are coming to kill us. Many animal studies scholars, such as animal trainer and philosopher Vicki Hearne (and likely lay observers of the intelligence of dogs), might counter that dogs in fact can do all these things, even regarding the execution of morally considered justice.[39] But rather than attempt to immunize against accusations of allegory with a defence drawn from realism, the sophisticated strategy of *White God* draws on cinematic renditions of revolution *and* gives loving cinematic attention to the bodies and the points of view (in both senses) of dogs *and*, in conjunction with the critical material presented in this chapter, enables further thought on the legal status of animals before the law and before justice.

Blood count

The suturing of the cinema and the slaughterhouse is neither new nor incidental. While this conjunction is perhaps immediately suggestive of horror genres, and the question that cannibal films such as *The Texas Chainsaw Massacre* pose regarding the difference between a house and a slaughterhouse – a difference usually directed at a moral or lawful question – deeper histories and economies have been brought to light in the work of Nicole Shukin.[40]

In her revised account of modernity, *Animal Capital: Rendering Life in Biopolitical Times*, Shukin showed how Henry Ford's well-known promulgation of the industrial assembly line was inspired by 'moving lines' achieving quite a different work: the 'disassembly lines' of the slaughterhouse.[41] With this mimetic inversion of assembly and disassembly in mind, Shukin connects her insight with the effect of animation produced by sequences of still images, in other words, cinema. As Noëlie Vialles observed in her ethnography of abattoirs, 'seeing round an abattoir in the opposite direction would be like watching a film backwards.'[42]

Cinema is not only bound up with a representational sleight of hand regarding the appearance and disappearance of the dead but also implicated in a material sense: the capacity to form a substance – photographic gelatin – that would 'fix' images was itself dependent on 'the waste of industrial slaughter'.[43] The insidious dependence of the moving image on rendered animal bodies leads to these two apparently discrete industries – cinema and agriculture – becoming increasingly and strategically enmeshed. Among the most astonishing histories unearthed by Shukin are the early-twentieth-century tours of modern American slaughterhouses couched as spectacles for white-middle-class family entertainment equivalent to a day at the fair. While the latter was also explicitly involved in a kind of pedagogy of affect – teaching audiences how to feel about what they saw with the aid of an illustrated guidebook in which a white little girl leads this instruction – Shukin's more recent work with Sarah O'Brien has examined the way that Sergei Eisenstein's 1925 film, *Strike*, appropriated the anticipated emotional response to footage of bovine slaughter in order to foment revolution.[44] One blow would produce another blow.[45] The irony is, as Shukin and O'Brien point out, that while Eisenstein wanted the death of animals to prompt the workers of the world to unite in revolution (the latter were supposed to take the former only allegorically), the workers too were being pummelled into action if 'only' by cinematic montage rather than by a spontaneous political uprising.[46]

Apart from the opening visual introduction to the father at work and his panicked attempt to find flame-throwers in the penultimate scenes, *White God* only films the exterior architecture of the slaughterhouse. Like other modern cities, Budapest has moved its slaughterhouses out of central locations, as meat districts become 'the Meat District', serving real estate development, and independent abattoirs become absorbed into anonymous animal industrial complexes. For the purposes of the film, *White God* insists on a central location within the flat historic streets of Pest: after Lili and Hagen are dropped off there at the beginning of the film, the sequence closes with a shot looking out of the courtyard. As cattle are led across this shot, the hills of Buda are clearly visible in the distance.[47] This is not because the film wants to make a point about the erstwhile skills and responsibility to individual animals of pre-industrial, small-scale, artisanal butchery but rather because it insists on the proximity of house to slaughterhouse.[48] The slaughterhouse is uncomfortably 'close to home'.

In Vialles' extraordinary anthropological study of slaughterhouses and those that work in them in the Adour region of France, she notes that the noun *'abattoir'* derives from the verb *'abattre'*, meaning 'to fell', in the sense of forestry, that is, felling a tree. Vialles remarks that the *Littré* dictionary translates it as 'to bring down that which is standing'.[49] Like subsequent work looking at industrial slaughter conducted in the United States, such as that of Charlie LeDuff and Timothy Pachirat, Vialles details the various systems, architectures, ideas and names that distance the workers from the actions that they perform. These systems issue tight guidelines and restricted sightlines, most especially to obscure the moment of actually killing animals, for which their exsanguination is central.[50] Vialles cites Judeo-Christian theological authority for and regulation of putting animals to death for human consumption. The book of Deuteronomy gives the imperative 'You shall not eat anything that dies of itself' and butchery practices translate this into a blood ordinance.[51] Vialles writes: 'Every animal must be killed by bleeding, and this must be done in an abattoir; these two conditions must be fulfilled for the meat to be deemed suitable for human consumption.'[52] Any abattoir must consequently plan its operations to take into account a lot of blood to be spilt and a lot of blood to be cleared away, whether through drains into running water or turned into animal food or agricultural products.[53] The intensive and systematic operation to drain blood away – to drain it from animal bodies and to drain it from the place of their slaughter as if it was never there – is typically reinforced by the newfound industrial insistence on the sterility of white. Vialles remarks: 'the colour of blood has been everywhere ousted by white: white walls, white accessories, white clothing from head to foot.'[54] A 'whitewashing' that is both literal and figurative aggressively blots out the sight of blood.

In the course of his seminars on the death penalty, Derrida asks, 'What is the *meaning* of cruelty?'[55] Referencing its Latin etymology, he continues:

> Is it *blood*, a history of blood, as the etymology seems to indicate (*cruor* is red blood, blood that flows)? And does one put an end to cruelty on the day that one no longer makes blood flow? Or else does cruelty point toward a radical evil, an evil for evil's sake, a suffering inflicted so as to make suffer, with or without blood?[56]

Blood is shed or shown only rarely over the duration of *White God*, even as the eugenic figures of 'mixed-blood' versus the 'pure' pervade this rendition of Budapest. The colour red is largely absent from the palette of the film, with the majority of its set design tonally blending with the black, grey, white and tan bodies of the dogs. Red, as Tobias Menely notes when writing on a 'prismatic ecology' beyond the more sedimented affirmative chromatic metonymy 'green', habitually 'signals disturbance and rupture'.[57] When it shows up in this 'animal horror' film, it is a bloodstain. The cinematic cut that articulates the divorcing family and the work of the slaughterhouse shows mere traces of blood on the floor – this underscores the cut and is part of its shock. It is a striking cinematic decision in the sense mobilized by Shukin and O'Brien above. But the bovine carcass, and its exposed heart, is exsanguinated, moderating its flow and muting its blow. Life has already been cleared away save for the single dot that still manages to stain the father's white shirt, still manages to mark the man whose role is ostensibly the furthest from the killing floor. In contrast, the promise of blood is realized during the scenes of Hagen's appropriation by the underground traffic in dogfighting; indeed, it saturates these scenes. The gypsy who uses Hagen – renamed 'Max' – in this way chooses him because his soft looks suggest that he 'still has a heart'. Having no interest in retaining this 'heart' – still less from 'learning by heart' – the gypsy covertly sharpens Max's teeth in a cruel technological insurance of and insistence upon nature as 'red in tooth and claw' (in Tennyson's famous phrase).[58] This covert operation also ensures that no one betting on this fight will speculate on the dog that still looks like a docile family pet, especially when he is pitched against a Rottweiler. It is a gamble that 'Max' will be taken for dead meat, a dead loss even before entering the arena.

In this place, blood flows by proxy as it falls prey to the pernicious substitution of beast and sovereign. Close-up shots show the two men opposite each other holding back their dogs until the ring is secured. Betting on the lives and deaths of each other's dogs, violence towards animals is always 'really' waged and wagered against our fellow man. It is blood money. Blood is spattered across the floor from the bodies of dead dogs dragged from the fighting ring, making way for Hagen to be led in.[59] Blood is never obscured or wiped away here – but the barriers that form the ring curtail it. (It should be noted that the film depicts what is nevertheless a sanitized version of the training of dogs for such fighting. That is, while it shows 'Max' being forced to fight for meat, being beaten so that he no longer trusts humans and being trained on a treadmill for endurance, it refrains from showing other animals being used as 'bait' or dwelling on what happens to dogs that fail to win, and it is cleverly cut so as to avoid showing any blows or bites actually striking their victims.)[60] Blood will be drawn in this space, this arena: it is to be seen and to be overseen by those men – there are only men in this economy – who are betting upon its flow. In being drawn, that is both made to flow and figured, blood is also the traffic between the sensible and the intelligible, as Gil Anidjar has suggested.[61] The architecture of the makeshift ring concentrates an idea of the raw – of raw nature, raw wounds, raw blood – even as the men organizing such events have cultured its emergence and its distance from themselves.

While the architectural sightlines within slaughterhouses are subject to deliberate planning, which is to say restriction on who can see what, Vialles suggests that the smell of blood is a lingering problem for human workers. This leads her to consider the heightened sense of olfaction in many mammals as compared to humans. She writes:

> It seems inconceivable that sensory perception should not induce signification. The fact that animals *sentent* ['perceive through the senses', but also literally 'smell'] necessarily means that they *pressentent* ['experience foreboding']. And discourse embarrasses itself when it seeks to separate the two [*sentir* and *pressentir*], so closely are they linked in common parlance, for which the sense of smell is the faculty of intuition.[62]

Figure 7.3 Blood (*White God*, 2014, Dir. Kornél Mundruczó, Hungary)

The matter or problem of smell for animals in slaughterhouses, in which their own imminent death hangs in the air, does not come under further consideration in Vialles' anthropology, leaving the question of both the recognition of death as such and the anticipation of death for these animals unexplored.[63] *White God* is at pains to suggest that dogs can anticipate their own death and thus would suffer in such anticipation. As suggested in the last section, Hagen escapes the dogfighting ring after he has been led to kill another dog, one blow leading to another. *White God* shows us his hesitation and realization of the death that he has caused in a reaction shot centred upon his bloody jaws. Hagen later leads the escape from the shelter after witnessing another dog being euthanized through a door held ajar and understanding that his time will be up next. Losing what is a brief reprieve from being euthanized by biting a potential adopter, Hagen is led away to rejoin the dogs destined for death. Resisting this end, he kills the man leading him. This time his bloodied jaws are seen by a cage full of dogs, who are clearly meant to understand this as illuminating and righteous resistance. The film thus retains a visual register for the apprehension of the death of another and hence one's own finitude and assigns this register to non-human animals. This register is by means of entirely conventional point-of-view shots (conventional except that they convey the point of view of a dog). The consequent emphasis on visual over conceivably olfactory cues is arguably complicit in a certain anthropomorphism. On the one hand, point-of-view shots are a central part of classic cinema's armature of identification. It is not just Hagen that sees the bleeding body of his opponent; it is not just Hagen that sees through a crack in the door that a dog is being put down: we also line up with his viewpoint. It is Hagen's point of view that we are encouraged to take as our own, not that of those orchestrating these events. On the other, so much attention to the visual remains within the discourse of sovereignty as both ocularcentric and anthropocentric. Following Derrida's analysis of an autopsy in *The Beast and the Sovereign*, structured by the visual survey of an elephant's exposed body by the Sun King, Louis XIV, Kelly Oliver remarks that 'man's dominion over other animals is built on an model of sovereignty as necropsy that erects itself through the autopsic model of power, which ultimately is built on the scaffolding of death and the death penalty'.[64] It is a sightline up to and including putting the other to death.

As the more palatable euphemism for what Derrida critically names a 'non-criminal putting to death' in distinction to murder outlawed by the legal and ethical command 'Thou shalt not kill', 'slaughter' is everywhere implied in a city rounding up those determined as having 'impure' blood. In 'The Animal Cure', I detailed the lethal slippage between the implicit performative address to a subject both human and fraternal and the consequent expendability of those non-fellows without due legal protection, drawing centrally on the interview with Derrida called 'Eating Well'. It is clear here, in the eugenically charged context of Budapest in *White God*, that those without legal – or ethical – protection are not *only* animals but those whom we *call* 'animal', issued death sentences with such figures as 'vermin' or 'swarms,' or even 'dogs' or even 'gypsies'.[65] In 'Eating Well' Derrida underscored the sacrificial aspect of metaphysics with the prefix 'carno-' adumbrating what he had previously termed 'phallogocentrism'. In the *Death Penalty* seminars, which took place some twelve years later, Derrida identifies the 'deconstruction of carno-phallogocentrism' as always attached to the 'deconstruction of this historical scaffolding of the death penalty'.[66] Symbolic law and state law coincide in their inscription of what is proper to man. Although Derrida does not engage the animal question at length in these seminars, nevertheless, he remarks that

> at the horizon ... obviously, there is the question of man's putting to death of animals and of whether one can speak of a death penalty inflicted by man on animals, or whether the death penalty is something proper to man, a putting to death only of man by man and not of one living being by another living being in general.[67]

Excavating historical instances of the implementation of the death penalty upon animals by humans, Oliver observes that 'from the Middle Ages through the seventeenth century with the first codifications of law, punishment was exacted against both man and beast alike to establish the scope of its sovereignty over all of creation'.[68] The history of capital punishment is thus bound up with that of the law and human treatment of other animals. Shukin points out the Cartesian contradiction in apparently assigning animals the capacity to wilfully transgress the law and thus be punished by it. She devotes considerable attention to the subject of Thomas Edison's early film *Electrocuting an Elephant* [1903], 'Topsy' the elephant, so named in a condensation of

American racism and speciesism (while the elephant was in fact Indian, this name merged her with African American slave labourers to whom it had been generically applied).[69] Having noted earlier Shukin's biopolitical account of the interwoven development of the cinema and agriculture, in this context, she further reveals the supplementary relationship between cinema and electricity as one of 'animal capital'.[70] Electrocuting an elephant, for Edison, Shukin suggests, 'would communicate once and for all the "actuality" of alternating current's deadliness to Westinghouse supporters [i.e. rivals in the emergent supply of electricity], its efficiency as a technology of death to penal authorities in New York State, and its painlessness to humane societies'.[71] The desirability of apparent painlessness vis-à-vis the maintenance of the death penalty as legally mandated is part of the same logic that ensures the blood flowing in slaughterhouses is diverted from human apprehension as much as possible. Indeed much recent attention to the death penalty – especially that carried out by lethal injection in many states of America – has been precisely to the issue of pain and the cocktail of drugs that attempt to anesthetize the body of the condemned prior to death, or at least to make it appear that the condemned does not suffer.[72] Paramount, finally, here is that pain does not visibly surface in the juridically authorized death penalty, the very idea of which becomes identified with an anaesthetic logic for Derrida.[73]

Derrida does not aim to consolidate the proper of man, still less to do so by means of the death penalty. Rather, as Elizabeth Rottenberg comments with a deliberately theatrical figure, he 'wants to *bring down the curtain* on the death penalty'.[74] This claim for the proper of man as the capacity to execute the death penalty is precisely, however, what he calls 'the classic philosopheme of all the great right-wing philosophies' (naming Kant and Hegel in particular).[75] All those philosophies that coalesce around the subject that calls himself man as that being who is raised up, above and beyond the animal that he once was (no following, not ever), pose law as more valuable than life, law as that which is entitled to sacrifice life, law as human prerogative. They pose the death penalty as a social contract.[76] Summarizing such philosophies, Derrida writes:

> The dignity of man, his sovereignty, the sign that he accedes to universal right and rises above animality is that he rises above biological life, puts his life in play in the law, risks his life and thus affirms his sovereignty as subject

or consciousness. A code of law that would refrain from inscribing the death penalty within it would not be a code of law: it would not be a human law, it would not be a law worthy of human dignity.[77]

White God focuses on the abattoir's open-sided courtyard: the place in which under 'normal' conditions, only humans may survive passage in both entry and exit. From the moment that the point of view of the father – or what becomes his erstwhile point of view – becomes the framing of the extended last shot, that framing looks downwards into the courtyard without a view or sense of an outside. The angle accentuates the sense of this space as an explicit condensation of many arenas – including that of a Roman amphitheatre. Surveyed from the perspective of the central exterior staircase of the slaughterhouse, the tilted, broadened view of the courtyard is revealed as an arena: it is a theatre; it is an auditorium; it is a courtroom. This angled view broadens the courtyard and narrows the road that leads in and out, rendering the insinuation of an erect phallus as the speculative ground on which the action takes place. There is a clear narrative expectation of a spectacular and, yes, climactic, battle. The time has come: this is what we have been waiting for. This is what we have been betting on. The stage has been set – and *White God* has ensured that there is a stage; there is yet another stage. The film's insistence upon spectacle exposes the virtual stage within the slaughterhouse, the minimal indicator of which lies in the official inspection of what is fit for human consumption, and extends it as a public desire that recalls the state's implication in the execution of the death penalty. As Derrida writes, 'The state must and wants to *see die* the condemned one.'[78] In this desire, the state

> best sees itself, that is, it acknowledges and becomes aware of its absolute sovereignty and that it *sees itself* in the sense in French where '*il se voit*' can mean 'it lets itself be seen' or 'it gives itself to be seen'… For this act of witnessing – the state as witness of the execution and witness of itself, of its own sovereignty, of its own almightiness – this act of witnessing must be visual: an eye witness. It thus never happens without a stage.[79]

The almighty quality of the state as witness, and its totalizing gambit, gives the lie to secularity. 'There is *theologico-political* wherever there is death penalty', Derrida argues.[80] Authority comes from on high in order to marshal a 'Thou shalt die'. That this stage and staging of death can appear to happen without blood returns us to the transubstantiation of sensible to intelligible, for Derrida finds that this scaffolding of the death penalty is bound not just to sovereignty and spectacle, and speculation as calculation, but it is also supported by 'speculative idealism'.[81] Carno-phallogocentrism thus must presume non-divisibility and spill not one drop.

Who stands before the law in this climactic arena? Who is the condemned? Up until the very last scene, the troubling resemblance of beast and sovereign hovers before us. If Hagen and his beast army are to die, then the 'natural order' of things resumes; if the humans are to die, then a reversal in this order will take place but headed by a new sovereign the order itself is maintained. On the one hand, there is a narrative expectation of execution, held within these poles of beast and sovereign, but on the other, there is a force of fascination at work in the staging of the death penalty that Derrida names a 'phantasm' of mastery.[82] This phantasm returns us not to the question of whether animals can recognize death, their own or that of another, but in the gesture that Derrida continually activates, shaking the grounds of thought in the process: whether we humans can do so. This phantasm of mastery exerts a fascination that seems to be realized by the death penalty. In terms that explicitly invoke cinema, he writes of this phantasm as that which

> we project ... as one projects a film ... we see in projection actually enacted what we are dreaming of all the time ... namely, to give ourselves death and to infinitize ourselves by giving ourselves death in a calculable, calculated, decidable fashion; and when I say 'we,' this means that in this dream we occupy, simultaneously or successively, all the positions, those of a judge, of judges, of the jury, of the executioner or the assistants, of the one condemned to death, of course, and the position of ones nearest and dearest, loved or hated, and that of the voyeuristic spectators who we are more than ever.[83]

This projected scene would eat everything in the name of 'giving ourselves death in a calculable fashion'. It would conscript all positions. This desire is tied to the worst of all machines. Not the Guillotine, not the electric chair and not lethal injection: the heart-stopping operations of the death penalty in all of its guises achieve their worst form in the timepiece that calls time. The clock and the calendar, the time and the date that re determined by the judge as metonymy of the sovereign state appears to master the demise of my own 'ticker' as that which is otherwise unmasterable, unknowable, unpredictable. Yet Derrida is clear that the death penalty as such cannot command the final materialization of the phantasm, clocking in on time. Rather, it can only remain phantasmatic and in excess of an approved cast. The attempt to calculate all in order to master all, and first and last of all, to master death,cannot ensure this desired infinitude: calculation replaces the future with the same. Calculation annihilates the future as such. In that event there would 'no longer [be] any event to come' even, Derrida says, 'no more heart of the other'.[84] Rather, he affirms:

> [o]nly a living being as a finite being can have a future, can be exposed to a future, to an incalculable and undecidable future that s/he does not have at his/her disposal like a master and that comes to him or to her from some other, from the heart of the other.[85]

Laying down the law

Most reviewers generally take a canine revolution as the film's 'must-see' conclusion. Some are disappointed that the revolution is not quite actualized according to its programme. Some get as far as Lili playing her trumpet to Hagen, raising a correct note of caution that music has 'tamed the beasts'. The film has contrasted its diegetic music – Lili rehearsing Liszt with her school orchestra – with the extra-diegetic grungy tune, to which beat the dogs run. In the slow confrontation that leads to the place of conclusion, intimately matching scale, angle and framing on the girl's feet back-tracking with the pack of dogs advancing, her previous exceptional status enters uncertain ground. A 'fan's' purely misogynist gesture circulates an alternative re-edited ending on YouTube that implies her grisly assault by this pack, but that is not an end that this film entertains. The pack is clearly being led by Hagen, who now refuses any special relationship with a human. Hagen is seen standing before Lily, his face redescribed into a lethal snarl by his artificially brutal teeth, so much so that she reneges on her earlier promise never to train him and pathetically attempts distraction with a game of 'fetch'.

'Fetch' fails for Lili and serves only to further align Hagen and his band of dogs against her. She plays a few phrases on her trumpet in a last-ditch effort to save herself from the pack and to save the pack from being attacked by her father. The crucial quiet moment that follows her musical half-measure has gone AWOL in the film's reception.[86] *White God* doesn't end there. It steps back from installing another sovereign, here in the form of the beast. The dogs now lie down around Lili, standing there as both performer for and conductor of this army-become-audience in the film's continued attention to what it is to stage. But her next gesture initiates the film's break with a predictable formula and the expectation of a final and fatal show of force: tearing up, Lili puts down her trumpet and lies down opposite Hagen, mirroring him. Their silent reappraisal of each other is shown via a classic shot-reverse shot sequence. Some ninety seconds follow before the credits roll. With the father having witnessed this transition, and perhaps also been affected by or calmed by his daughter's music for the first time, *White God* delivers a *coup de théâtre*. It allows the father to put off the slaughterhouse worker's call to the police. Cinematic renditions of the scene of state-sponsored execution frequently make use of a telephone call to heighten the suspense as to the delivery, or not, of the fatal blow (whether of a trapdoor or other trigger). 'There is always a telephone today', Derrida remarks, 'that links, like an umbilical cord of life or death, the place of execution to the executive power of the sovereign … who can grant a pardon'.[87] The telephone in the execution chamber usually connects a power to stay execution, but the telephone in *White God* threatens the reverse. Phoning the police in this context will reach a force that is already primed and has already promised to kill these dogs before the morning (a promise made to Lili). As with *White Dog*, the law in its embodied force is deferred, but in *White God*, a pact does not re-consolidate around a carno-phallogocentric economy. Like Lili's seemingly minor gesture of putting the trumpet down, the father merely halts a phone call. Where she arrests the narrative of blood vengeance and then, at first, supplants it with music before ceasing this gentler form of temporization too, he simply says: 'Give them a little more time'. We do not know how much time, only that in this fraught moment, when time is of the essence, in the face of the imminent destruction of life that would evacuate any revolutionary potential and recoup the slaughterhouse grounds as the place where animals come to die, they are to be given 'a little more'. Open-ended, this gift annuls the equivalence that *jus talionis* aims to balance. In this offering, the father effectively disconnects his metonymical alignment with the '*chef d'etat*' and opens the possibility of a pardon himself. 'The pardon [*la grâce*]', as

Derrida remarks, 'gives time, and the only "thing" that can be given graciously is time, that is to say, at once nothing and everything'.[88] Continuing with its radical revision of narrative closure, in *White God* the father does not rise to a position of sovereignty, or engineer another to do so in his stead. The title of sovereign falls away. This temporal donation returns the living to their proper finitude, one that cannot be calculated.

This donation constitutes the last dialogue in the film (unlike in the trailer, Lili does not say, 'I love you too' to Hagen, does not domesticate the scene with a pat response).[89] The communication between all figures in this place now becomes entirely gestural. While Lili and Hagen remain central to the shot, the father joins them. He steps down from his position and from the elevated balcony, descending in all senses to the scene within this courtyard, no longer above it. Walking over to Lili, he lies down to one side taking his lead from her. It is quiet – but not silent. There is a faint extra-diegetic classical sound of brass instruments, but it is the gentle diegetic sound of bird song that now animates this space, implying the dawn. Dawn as an index of the clock, that is, the worst machine. Dawn with its legal temporal signature of the arrival of the decision par excellence: execution or pardon of the death sentence. Dawn, with its double sense of the diegetic police, promises that the dogs would be dead by morning but also an extradiegetic sense of a hopeful new day. Dawn is signalled diegetically within this final scene, but its chorus continues across the closing credits: the birds keep singing. Thus their song maintains this spaced-out extension of the dawn and participates in the deferral of the law. Here in the place of death, lying down but not dead, human verticality has been vacated. In this revised orchestration of scene and protagonists, who is before whom?

Who is before what? The question leads straight back to the difference between – the *decision* between – a 'who' and a 'what' and the death sentence differentiating murder and slaughter. Are the dogs the revolutionary agents rushing in to destroy the conditions that create their class, that create two great classes: the 'who' and the 'what'? In this wilful horizontality of the living assembled in this arena, with vulnerability in common and hesitation suspending the order of things, the revolutionary gesture is not the violent overcoming of one class by another. Rather, a 'force of weakness' disorients their distinction. This is the heart of the matter, this alteration in address at the end. Address, here, encompasses the location of the slaughterhouse and all the autopsic architectures that I have suggested *White God* cinematically convenes. It is also the destination slipping into *destinerrance*, fostering address as a cinematic form of apostrophe when the film slips away from conventional narrative resolution and into a poetic aperture. It is this which returns us to the heart: '[t]he poetic, let us say it, would be that which you desire to learn, but from and of the other, thanks to the other and under dictation, by heart; *imparare a memoria*.'[90] In the *Death Penalty* seminars Derrida characterizes the heart of the other as under threat from cancellation by the automatic ticker: the clock and its state emissaries. This heart is not confined to the cardiovascular but beats too within a deconstructive psychoanalytic register such that

> the other whose heart is more interior to my heart than my heart itself, which means that I protect my heart, I protest [against the death penalty] in the name of my heart when I fight [*en me battant*] so that the heart of the other will continue to beat [*battre*] – in me before me, after me, or even without me.[91]

When Derrida writes of poetry or of what he insists on calling the 'poematic' in '*Che cos'è la poesia?*' two aspects of his enigmatic little text resound here. One is the animal figure 'turned toward the other and toward itself', the humble hedgehog that remains 'close to the earth' blind to the arrival of death.[92] The other is the invocatory structure that he gives to the poematic, which is that of learning by heart. It is a love that departs from 'I love you too'. The heart joins the two ostensibly radically different texts. Confounding any inherited sense of what may ordinarily be thought of as a poem or any sense of an autopoesis, Derrida's cardio-pedagogy is both addressed to you and signatory of you, we might even say that it must *carry* you (to echo Derrida's writing on Celan).[93] Eschewing particular histories of form, this poematic renders an event that strikes up a conversation with the iterability characteristic of Derrida's more well-known work on the performative. What cannot be missed in '*Che cos'è la poesia?*' is the invocation of love, of passion and the heart, not because of but inextricable from its second-person address to *you (tu)*. Even if Derrida also describes this poem as a 'dictation' since it demands to be learnt by heart, there is no sense in which this could be absolutely performed or retained – the vulnerability of the poematic hedgehog sees to that.[94] Instead, he says, '[t]his "demon of the heart" never gathers itself together; rather it loses itself and gets off the track ... it exposes itself to chance, it would rather let itself be torn to pieces by what bears down upon it.'[95]

In 'Force of Law' Derrida writes of the violence inherent to law, of the interweaving of violence that founds and violence that conserves the law, of the rote citation of the law ghosted by the possibility of a wholly new decision that might do justice to the singularity of the other, of the terrifying experience of non-law in law in the revolutionary moment of founding another:

> It is the moment in which the foundation of law remains suspended in the void or over the abyss, suspended by a pure performative act that would not have to answer to or before anyone ... The supposed subject of this pure performative would no longer be before the law, or rather he would be before a law still undetermined, before the law, as before a law still nonexisting, a law still ahead, still having to and yet to come.[96]

In *The Beast and The Sovereign*, Derrida continues this thought such that its utter transposition of 'our' condition is unmistakable. Rather than the grand scene of history, law, culture and representation arising out of the primal patricide that will ever be repeated:

> The 'unrecognizable'... is the beginning of ethics, of the Law, and not of the human ... So long as it remains human, among men, ethics remains dogmatic, narcissistic, and not yet thinking.'[97]

Figure 7.4 End (*White God*, 2014, Dir. Kornél Mundruczó, Hungary)

In the hiatus of the film's last minute, the heart beats faster. The end is upon us, but it is not what was expected. Close to the earth, exposed to chance, laying down before a law not yet existing, what remains of the subject? In the abdication of paternal authority, the audience is the only other site of appeal. The last, static, shot remains elevated, looking down upon those eschewing the rectitude, the moral 'uprightness', of the law. In this arena all bets are off: 'our' dogs are not fighting anymore. Still shepherding the shadows of autopsic architectures, attention is now on you.

This cinematic apostrophe addresses you. What will you do? You could step down and join them: there's plenty of room.

Notes

Chapter 1

1. Jacques Derrida, *The Animal That Therefore I Am*, trans. David Wills, ed. Marie-Louise Mallet (New York: Fordham University Press, 2008) 1.
2. Timothy Clark, 'By Heart: A Reading of Derrida's "*Che cos'è la poesia?*" Through Keats and Celan' in *Oxford Literary Review*, 15:1 (2012): 72, n.1.
3. See Clark, 'By Heart', 43.
4. Jacques Derrida, '*Che cos'è la poesia?*' [1988] in Elizabeth Weber, ed. *Points ... Interviews 1974–1994* (Stanford: Stanford University Press) 291.
5. Jacques Derrida, '*Istrice2: Ick bünn all hier?*' in Elizabeth Weber, ed. *Points ... Interviews 1974–1994* (Stanford: Stanford University Press) 302. Schlegel's fragment – compared to a hedgehog – is entirely isolated and cut off from the world.
6. Derrida, '*Che cos'è la poesia?*' 295.
7. Ibid.
8. Derrida, qtd. in Derrida, *Animal*, 36.
9. Jacques Derrida, 'A Silkworm of One's Own' in Hélène Cixous and Jacques Derrida, eds. *Veils*, trans. Geoff Bennington (Palo Alto: Stanford University Press, 2001) 17–92. Compare the classical account in Sigmund Freud, 'Fetishism' [1927] in *On Sexuality, V.7 in the Freud Library* (London: Penguin, 1991) 345–58.
10. Derrida, *Animal*, 37.
11. Jacques Derrida, 'Women in the Beehive: A Seminar with Jacques Derrida' [1984] in Alice Jardine and Paul Smith, eds. *Men in Feminism* (London: Methuen, 1987) 196. Described as 'authorised but authorless', the seminar depended on the editors' transcription of Derrida's remarks, and those editors are unnamed.
12. Jacques Derrida, *The Death Penalty V. I*, trans. Peggy Kamuf (Chicago: Chicago University Press, 2014) 257.
13. See Ranjanna Khanna, *Dark Continents: Psychoanalysis and Colonialism* (Durham: Duke University Press, 2003).
14. Sigmund Freud, 'The Taboo of Virginity' [1917] in *On Sexuality, V.7 in the Freud Library* (London: Penguin, 1991) 279.

15 Freud, 'Virginity', 279, the speculation is with approving reference to the work of Sandor Ferenczi.
16 Jacques Derrida, *The Death Penalty V. II*, trans. Elizabeth Rottenberg (Chicago: Chicago University Press, 2017) 232.
17 Derrida, *Death Penalty II*, 232–3.
18 Derrida, *Animal*, 28.
19 Donna J. Haraway, '"Gender" for a Marxist Dictionary' [1987] in *Simians, Cyborgs & Women: The Reinvention of Nature* (London: Free Association Books, 1991) 129.

Chapter 2

1 Jacques Derrida, 'Mnemosyne', trans. Cecile Lindsay in *Memoires for Paul de Man*, revised edition (New York: Columbia University Press, 1989) 31.
2 *My Talks with Dean Spanley* (2008) [Film] Dir. Toa Fraser (UK/NZ: Icon Film). *Dean Spanley* is now re-published as one volume containing both Lord Dunsany's novella *Dean Spanley* and Alan Sharp's screenplay, ed. Matthew Metcalfe with Chris Smith (London: HarperCollins Publishers, 2008). The film trailer is accessible here: https://www.imdb.com/title/tt1135968/?ref_=fn_al_tt_1. Accessed 22 May 2020.
3 See Tom Tyler, '*Quia Ego Nominor Leo*: Barthes, Stereotypes, and Aesop's Animals' in *Mosaic*, 40:1 (2007): 45–59.
4 One of the first lessons of Jacques Derrida's work on 'the animal question' is that the definite article performs a work of capture designed to police a firm boundary between the human and the animal. See his *The Animal That Therefore I Am*, trans. David Wills, ed. Marie-Louise Mallet (New York: Fordham University Press, 2008).
5 I discuss the vexed relation between 'trauma studies' and 'animal studies' in 'Voice' in Lynn Turner, Undine Sellbach and Ron Broglio, eds. *The Edinburgh Companion to Animal Studies* (Edinburgh: Edinburgh University Press, 2018) 518–32.
6 See Sigmund Freud and Josef Breuer, 'On the Psychical Mechanism of Hysterical Phenomena: Preliminary Communication' [1893] in *Studies on Hysteria* No. 3 in the Penguin Freud Library, trans. James Strachey (London: Penguin Books, 1991) 53–72.
7 Hunting for more Tokay, Wrather takes Fisk to the Nawab, who coincidently refers to the Dean as Old Wag Spanley, referring to the name he was known by at Oxford by virtue of his initials (Walther Arthur Graham Spanley).

8   Sigmund Freud, 'Totem and Taboo: Some Correspondences Between the Psychical Lives of Savages and Neurotics' [1913] in *On Murder, Mourning and Melancholia*, trans. Shaun Whiteside (London: Penguin Books, 2005) 1–166.
9   See Ranjanna Khanna, *Dark Continents: Psychoanalysis and Colonialism* (Durham: Duke University Press, 2003). There she suggests that 'psychoanalysis and ethnology participate in the same episteme, one that sustains, through calm violence, the sovereign subject of Europe' (68).
10 For one of the most helpful commentaries, see Leonard Lawlor, *This Is Not Sufficient: An Essay on Animality and Human Nature in Derrida* (New York: Columbia University Press, 2007).
11 Derrida, *Animal*, 36.
12 Derrida references numerous examples in *Animal* 36–8.
13 See Jacques Derrida, *Politics of Friendship*, trans. George Collins (London: Verso, 2005).
14 See Freud, 'Totem & Taboo', 140–3.
15 Ibid., 128, emphasis added.
16 Freud, 'Totem & Taboo', 141. Richard J. Smith strongly states that Darwin himself would not recognize what Freud writes in his name; see 'Darwin, Freud and the continuing misrepresentation of the Primal Horde' in *Current Anthropology*, 57:6 (2016): 838–43. Freud's gesture is fascinating – to call upon the 'father of evolution' to supply an origin that is not there.
17 See Kelly Oliver, *Animal Lessons: How They Teach Us to Be Human* (New York: Columbia University Press, 2009) 248–57 and Elissa Marder, 'The Sex of Death and the Maternal Crypt' in *Parallax*, 15:1 (2009): 5–20.
18 See Deborah Bird Rose, 'Totemism, Regions, and Co-management in Aboriginal Australia' conference paper at Crossing Boundaries, the Seventh Biennial Conference of the International Association for the Study of Common Property, Vancouver, B.C. 1998. http://dlc.dlib.indiana.edu/dlc/bitstream/handle/10535/1187/rose.pdf. Accessed 20 May 2020.
19 See Julia Kristeva, *Powers of Horror: An Essay on Abjection*, trans. Leon Roudiez. (New York: Columbia University Press, 1982) 56–89.
20 Kristeva, *Powers*, 12–13.
21 Ibid., 61, 63.
22 The narrative of reincarnation might itself be understood as a topic invested in minimising maternity.
23 Kristeva, *Powers*, 75.
24 Ibid., 102.

25 Ibid.
26 Sharp, 193–4.
27 Compare the very different intimacy between urination, menopausal women and equine hormones in Donna J. Haraway 'Awash in Urine: DES and Premarin in Multispecies Response-ability' in *Staying with the Trouble: Making Kin in the Chthulucene* (Durham: Duke University Press, 2016) 104–16.
28 The Dean says as much later in the film, during one of the 'animalseances'.
29 With so many doubles structuring this film together with the uncertainty regarding the veracity of the Dean's memories and the status of reincarnation, Freud's 'uncanny' is strongly evoked. Sigmund Freud, 'The Uncanny' in *Art and Literature*, V. 14 of The Penguin Freud Library, trans. James Strachey (London: Penguin, 1985) 336–76.
30 See Akira Lippit, *Electric Animal: Toward a Rhetoric of Wildlife* (Minneapolis: Minnesota University Press, 2000), 123. It is ambiguous as to whether Lippit affirms this view.
31 Sigmund Freud, 'Civilization and Its Discontents' [1929] in the Penguin Freud Library V.12 *Civilization, Society & Religion*, trans. James Strachey (London: Penguin, 1991) 288–9, n.1.
32 Maud Ellmann links Freud's theory of the human repression of olfaction with his own relationship with Wilhelm Fliess and the latter's investment in the nasal aetiology of hysteria; see her 'Noses and Monotheism' in Maureen O'Connor, ed. *Back to the Future of Irish Studies* (Oxford: Peter Lang Publishing, 2011) 165–76.
33 Freud, 'Civilization', 289, n.1.
34 Jacques Derrida, '"Eating Well" or the Calculation of the Subject,' in Elizabeth Weber, ed. *Points … Interviews 1974–1994* (Stanford: Stanford University Press, 1995) 282.
35 Sara Guyer, 'Albeit Eating: Towards an Ethics of Cannibalism' in *Angelaki*, 2:1 (1997): 64.
36 Derrida, 'Eating Well', 282. Originally published in *Topoi*, 7:2 (1988), 113–21.
37 Derrida, 'Eating Well', 282.
38 Prior to 'Eating Well', Derrida published 'Foreword: Fors: The Anglish Words of Nicolas Abraham and Maria Torok' [1977] in *The Wolf Man's Magic Word: A Cryptonomy*, trans. Nicholas Rand (Minneapolis: Minnesota University Press, 1986) xi–xlviii. '*Fors*' was thus originally published only two years after the publication of Derrida's first major critical engagement with Jacques Lacan, 'Le Facteur de la Verité'[1975] in *The Post Card: From Socrates to Freud and Beyond*, trans. Alan Bass (Chicago: Chicago University Press, 1987) 411–98. Scholarship

has yet to grapple with the relations between these colossal essays, though that might begin with the long and pointed parenthesis in which Derrida addresses Lacan's reputed repudiation of '*Fors*'. See Jacques Derrida, *The Beast and the Sovereign V. 1*, trans. Geoffrey Bennington (Chicago: Chicago University Press, 2009) 145.

39   Derrida, 'Eating Well', 282.
40   Ibid., 280.
41   See Lynn Turner, 'Insect Asides' in Lynn Turner, ed. *The Animal Question in Deconstruction* (Edinburgh: Edinburgh University Press, 2013) 54–69.
42   Derrida, 'Eating Well', 277.
43   Ibid., 280. Levinas's inability to think the 'face' as anything other than human has been the topic of extensive debate, e.g. John Llewelyn 'Am I Obsessed by Bobby? (Humanism of the Other Animal)' in Robert Bernasconi, eds. *Re-Reading Levinas,* or Deborah Bird Rose 'Bobby's Face, My Love' in *Wild Dog Dreaming: Love and Extinction* (Charlottesville: Virginia University Press, 2011) 29–41.
44   Remarkably, Italian feminist philosopher Adriana Cavarero's recent exegesis on the theological commandment and its Levinasian afterlife, 'Thou shalt not kill,' first published in Italian in 2011, includes only one brief reference to Derrida (to *Politics of Friendship* and the fraternal), a scholar that she lauded as a primary influence in her *For More than One Voice* (trans. Paul A. Kottman [Stanford University Press, 2005]). More remarkably, she makes only one oblique reference to an unnamed 'some' who 'now hold that the prohibition against killing may extend even to animals and take into account the biblical question about their slaughter, when it comes to *homicidium* it is only man (*homo*) that is spoken of'. See her 'The Archaeology of Homicide' in Adriana Cavarero and Angelo Scola, eds. *Thou Shalt Not Kill: A Political and Theological Dialogue*, trans. Margaret Adams Groesbeck and Adam Sitze (New York: Fordham University Press, 2015) 79. Otherwise, she completely ignores the ethical question of non-human others.
45   Derrida, 'Eating Well', 281.
46   Ibid., 282.
47   Discussed further in Lynn Turner '*Hors d'Oeuvre*: Some Footnotes to the Spurs of Dorothy Cross' in *Parallax*, 19:1 (2013): 3–11.
48   Melanie Klein, 'Weaning' [1936] in *Love, Guilt and Reparation and Other Works 1921–1945* (London: Vintage, 1998) 291, emphasis added. See also Lynn Turner, 'Fort Spa: In at the Deep End with Derrida and Ferenczi' in Simon Morgan Wortham and Chiara Alfano, eds. *Desire in Ashes: Deconstruction, Psychoanalysis, Philosophy* (London: Bloomsbury, 2015) 103–20.

49 Derrida, 'Freud and the Scene of Writing' in *Writing and Difference*, trans. Alan Bass (London: Routledge Kegan Paul, 1978) 231.
50 Sara Guyer elaborates the demotion of the mouth in 'Buccality' in Gabrielle Schwab, ed. *Derrida, Deleuze, Psychoanalysis* (New York: Columbia University Press, 2007) 80–1.
51 Derrida, 'Eating Well', 282.
52 Nicolas Abraham and Maria Torok, 'Mourning *or* Melancholia: Introjection *versus* Incorporation' [1972] in *The Shell and The Kernel*, trans. Nicholas T. Rand (Chicago: Chicago University Press, 1994) 125–38.
53 Derrida, '*Fors*', xxxvii.
54 Ibid., emphasis original.
55 Ibid., xxxvii, xxxviii, emphases original.
56 Derrida re-describes this demetaphorization as a 'hypermetaphorization'; see '*Fors*' xxxviii.
57 Ibid., xxxviii, emphasis original.
58 Abraham and Torok, 'Mourning *or* Melancholia', 128.
59 Derrida, '*Fors*', xvi
60 Ibid., xxxix.
61 See also Vicki Kirby 'Original Science: Nature Deconstructing Itself' in *Derrida Today*, 3:2 (2010): 201–20.
62 The script makes the direct analogy (243), deleted but deducible in the film itself.
63 Dunsany, 247.
64 See Derrida, *Animal*, 4, Lawlor, *This Is Not Sufficient*, 135, n.6.
65 Erica Fudge argues that a prime function of the pet is to 'come home'; see *Pets* (Stocksfield: Ashgate, 2008).
66 For Derrida's critique of the difference between the proper death of *Dasein* and the perishing of the animal in Heidegger, see *Aporias*, trans. Thomas Dutoit (Stanford: Stanford University Press, 1993) 30–1. I elaborate on this question in the final chapter of this volume.
67 Sharp's script has Fisk also mention Wag to Mrs Brimley: this is edited out in the film, 267.
68 In spite of Fisk's earlier protestations that he could never hope to have another dog like Wag – one of the 'seven great dogs' alive at any time according to his idiosyncratic mythology.
69 Henslowe showed his own distress over her death earlier in the film.
70 Rather more jovial than Fisk, Mrs Brimley is not unlike him in her literality: when someone dies, that is all there is to it; she may be talking to a chair but it was apparently just like that when her quiet husband was alive.

71 Derrida, *Animal*, 12, and Derrida, *Politics of Friendship*, 149.
72 Laurence Rickels, 'Pet Grief' in *gorillagorillagorilla*, Diana Thater ex. cat. (Köln: Walter König, 2009) 71.
73 Rickels, 'Pet Grief', 72.
74 Ibid.
75 Derrida, *Animal*, 8.
76 Kelly Oliver reverses the stakes and asks after the now established ability of elephants to mourn in 'Elephant Eulogy' in Lynn Turner, ed. *The Animal Question in Deconstruction* (Edinburgh: Edinburgh University Press, 2013) 89–104.
77 Derrida, *Beast 1*, 294.
78 See Derrida, *Beast 1*, 280–94. Incidentally, while Derrida does not know either the source or the manner of death of '*la bête*' in question here, Dawne McCance unearths the female sex of this elephant, who died aged seventeen in Versailles, having been gifted from Portugal to France in a further exercise of sovereign power. See Dawne McCance, *The Reproduction of Life Death: Derrida's La vie la mort* (New York: Fordham University Press, 2019) 99.
79 Derrida, *Beast 1*, 299.
80 Derrida, *Animal*, 27

Chapter 3

1 Julia Kristeva, *Powers of Horror: An Essay on Abjection*, trans. Leon Roudiez. (New York: Columbia University Press, 1982) 12–13, italics original. The apparent quotation is not noted in the text.
2 Julia Kristeva, 'The Impudence of Uttering: Mother Tongue', trans. Patricia Vieira and Michael Marder, in *Psychoanalytic Review*, 97:4 (2010): 679.
3 Kelly Oliver, *Animal Lessons: How They Teach Us to Be Human* (New York: Columbia University Press, 2009).
4 Oliver is addressing continental philosophy, not eco-feminism that explicitly invokes intersectional analysis. For the recent state of play in eco-feminism, see Carol J. Adams and Lori Gruen, eds. *Ecofeminism: Feminist Intersections with Other Animals and the Earth* (London: Bloomsbury, 2014).
5 See Lynn Turner, 'Animal and Sexual Differences: Kelly Oliver's Continental Bestiary' in *Body and Society*, 19:4 (2013): 120–33.
6 Beauvoir, qtd. in Oliver 165.
7 *Abject Art: Repulsion and Desire in American Art*, Whitney Museum of Contemporary Art, 1983. *Rites of Passage*, The Tate Gallery, London, 1995.

8   Kristeva 'Impudence of Uttering', 679.
9   Kristeva does not appear to have directed her attention to 'the animal' or animality as such at all. For a rare appropriation of her work in the context of animal studies, see Ruth Lipschitz, 'Abjection' in Lynn Turner, Undine Sellbach and Ron Broglio, eds. *The Edinburgh Companion to Animal Studies* (Edinburgh: Edinburgh University Press, 2018) 13–29.
10 Kristeva's defence of the singularity of language as human is given an especially tendentious assertion when she remarks that this is the case 'even in the most handicapped speaking subject' ('Uttering' 685). This implicitly counters the right's based discourse of utilitarian philosopher Peter Singer, tendentious in his own way regarding the capacities of nonhuman animals vis-à-vis humans with disabilities – acutely discussed in Dawne McCance, ed. *Critical Animal Studies: An Introduction* (State University of New York Press, 2013).
11 Famously, Kristeva's earlier forays into what she names the feminine were all vehicled by male modernist poets, such as Mallarmé and Celan. In 2001–2003 Kristeva published three books with Columbia University Press under the title of *Feminine Genius (Le Génie Féminin)*, and these were devoted to Hannah Arendt, Colette and Melanie Klein, respectively.
12 Kristeva, 'Uttering', 686.
13 See Barbara Creed and Jeanette Hoorn, 'Animals, Art, Abjection' in Rina Arya and Nicholas Chare, eds. *Abject Visions: Powers of Horror in Art and Visual Culture* (Manchester: Manchester University Press, 2016), 90–104, and Kelly Oliver, *Technologies of Life and Death: From Cloning to Capital Punishment* (New York: Fordham University Press, 2013). Oliver notes that when Kristeva subsequently returned to 'Totem and Taboo' in her work on revolt, the animal has almost entirely disappeared from her account. See Oliver, 298, and Julia Kristeva, *The Sense and Non-sense of Revolt* (New York: Columbia University Press, 2000).
14 See Christopher Powici, 'Who Are the Bandar-log?' Questioning Animals in Rudyard Kiplaing's Mowgli Stories and Ursula Le Guin's 'Buffalo Gals, Wont You Come Out Tonight?' in Mary Sanders Pollock and Catherine Rainwater, eds. *Figuring Animals: Essays on Animal Images in Art, Literature, Philosophy and Popular Culture* (New York: Palgrave, 2005) 179. Curiously, while Powici reads Kristeva widely – and notably *Revolution in Poetic Language* – he does not cite *Powers of Horror* at all.
15 Powici, 'Who Are the Bandar-log?' 179.
16 Derrida, *Animal*, 7.

17 What will be the biggest in a series of mass extinction events since the one that devastated the dinosaurs is now in train (made visceral by the devastating fires in Australia in the winter of 2019), see Elizabeth Kolbert, *The Sixth Extinction: An Unnatural History* (Bloomsbury Paperbacks, 2015).

18 Sigmund Freud, 'Civilization and Its Discontents' [1929] in the Penguin Freud Library V.12 *Civilization, Society & Religion*, trans. James Strachey (London: Penguin, 1991) 289, n.1.

19 *The Woman* (2011), [Film] Dir. Lucky Mckee, USA: Moderncinè, trailer accessible here: https://www.imdb.com/title/tt1714208/?ref_=fn_al_tt_1
The film loosely follows a previous horror film about a cannibal family, the last remaining member of which is the Woman – *Offspring* (2009) [Film] Dir. Andrew van den Houten, USA: Moderncinè, written by Jack Ketchum. At the time of writing, a sequel to *The Woman* had just been screened in film festivals: *Darlin',* directed by Pollyanna McIntosh (co-written by McIntosh and Jack Ketchum), USA: Hood River Entertainment, 2019.

20 See Carol J. Clover, *Men, Women and Chainsaws: Gender in the Modern Horror Film* (Princeton: Princeton University Press, 1992).

21 Cary Wolfe with Jonathan Elmer, 'Subject to Sacrifice: Ideology, Sacrifice and the Discourse of Species in Jonathan Demme's The Silence of the Lambs' in Cary Wolfe, ed. *Animal Rites: American Culture, the Discourse of Species and Posthumanist Theory* (Chicago University Press, 2003) 116.

22 Barbara Creed, [1993] *The Monstrous Feminine: Film, Feminism, Psychoanalysis* (New York: Routledge, 2007).

23 Pollyanna McIntosh gave a considered interview on sexual violence in cinema: https://podcasts.apple.com/us/podcast/pollyanna-mcintosh-and-staci-layne-wilson/id1453745253?i=1000432164016&mt=2

24 The Sundance film festival screening of the film infamously drew hostile responses from some male audience members: https://www.youtube.com/watch?v=o3lUAZLB4JY

25 Penn Bullock, 'Transcript: Donald Trump's Taped Comments about Women' *The New York Times*, 8 October 2016. https://www.nytimes.com/2016/10/08/us/donald-trump-tape-transcript.html. Accessed 15 February 2017. For a fuller discussion of the 'domesticated fetish of the term pussy', see our editor's 'Introduction' in, Turner, Sellbach and Broglio, eds. *Edinburgh Companion to Animal Studies* 6.

26 See Sabrina Siddiqui and Lauren Gambino, 'Brett Kavanaugh's Confirmation to Supreme Court Gives Trump Major Victory' *The Guardian*, 6 October 2018.

https://www.theguardian.com/us-news/2018/oct/06/brett-kavanaugh-confirmed-us-supreme-court. Accessed 22 February 2019.

27 For example, see Elizabeth Cowie's collection of writings from the 1970s to the 1990s: *Representing the Woman: Cinema and Psychoanalysis* (Minneapolis: Minnesota University Press, 1996).

28 Kristeva, *Powers*, 63.

29 Angela Bettis has frequently taken roles in horror films, often roles that exploited her slight frame for the surprise of supernatural or other powers. She played the title role in the 2002 television remake of *Carrie* and also starred in Lucky Mckee's earlier cult horror film, *May* (2002) USA: 2 Lop films. *May*, intriguingly, narrates precisely the confusion between literal and symbolic anthropophagy. This is hilariously articulated when Bettis' character has an unfortunate date with an artist making 'statements' about love as cannibalism only to encounter someone who cannot demarcate between a bite and a kiss.

30 Mark Wigley precisely links metaphysics and the home economy; see his *The Architecture of Deconstruction: Derrida's Haunt* (Harvard: MIT Press, 1995).

31 Wolfe with Elmer, 'Subject to Sacrifice', 101.

32 See Donna J. Haraway, *When Species Meet* (Minneapolis: Minnesota University Press, 2008) 17.

33 The novel does name the dogs and specifies that there are four. See Jack Ketchum and Lucky McKee, *The Woman* (Gauntlett Press, 2016).

34 The novel describes Darlin' as 'making believe' that she is the 'animal woman' and as such strong enough to chase down and eat men. Ketchum and McKee, *The Woman* 96.

35 Freud, 'Civilization', 293.

36 Ibid., emphasis added.

37 Ibid., 289, n.1.

38 The colonial imaginary of soap is detailed in Anne McClintock, *Imperial Leather: Race, Gender and Sexuality in the Colonial Contest* (New York and London: Routledge, 1995).

39 Kristeva, *Powers*, 102. Ruth Lipschitz gives an excellent account of the concatenated abjection of race, gender and species in her 'Skin/ned politics: Species Discourse and the Limits of "The Human" in Nandipha Mntambo's Art' in *Hypatia: A Journal of Feminist Philosophy*, 27:3 (2012): 546–66.

40 Sigmund Freud (from his essay 'The Question of Lay Analysis'), qtd. in Ranjanna Khanna, *Dark Continents: Psychoanalysis and Colonialism* (Durham: Duke University Press, 2003) 47.

41 Khanna, *Dark Continents*, 49.
42 Ibid., 52. Kristeva too remarks on the 'dark continent', even providing an entry for this phrase in Alain de Mijolla, ed. *International Dictionary of Psychoanalysis, Volume 1 A-F* (USA: Macmillan Reference, 2005) 365. She voices criticism of Freud's inability to theorise female sexuality and notes the 'murky' quality of this phrase without, however, acknowledging the colonial component at work. Kristeva's genealogy of subsequent work on female navigation of Oedipus concludes with her own work.
43 During the years I taught this material, several cases were widely reported regarding the actual long-term abduction of (white) women, typically within secret cellars in domestic houses. See for example Kirwan Randhawa and David Gardner, 'Ohio Dungeon: Three Women Rescued from Chains in Basement Prison after Ten Years' *Evening Standard*, 7 May 2013. https://www.standard.co.uk/news/world/ohio-dungeon-three-women-rescued-from-chains-in-basement-prison-after-ten-years-8605438.html and Mark Landler 'Austria Stunned by Case of Imprisoned Woman' *NYTimes*, 29April 2008. https://www.nytimes.com/2008/04/29/world/europe/29austria.html
44 Gayle Rubin, 'The "Traffic" in Women' in R. Reiter, ed. *Toward an Anthropology of Women* (New York: Monthly Review Press, 1975) 33–65.
45 Rubin, 'Traffic', 26.
46 Donna J. Haraway, '"Gender" for a Marxist Dictionary' [1987] in *Simians, Cyborgs & Women: The Reinvention of Nature* (London: Free Association Books, 1991) 130. I note the work of Derrida on the complexity of 'Geschlecht' in his several essays on this term in the work of Heidegger, also written in the 1980s in my 'Critical Companions: Derrida, Haraway and Other Animals' in Julian Wolfreys, ed. *Introducing Criticism in the 21st Century*, 2nd Edition (Edinburgh: Edinburgh University Press, 2015) 63–80.
47 The works cited are: Hazel V. Carby, *Reconstructing Womanhood: The Emergence of the Afro-American Woman Novel* (New York: Oxford University Press, 1987) and, Hortense Spillers 'Mama's Baby, Papa's Maybe: An American Grammar Book' *Diacritics*, 7:2 (1987): 65–81. Both references, along with the commissioning and first publication of Haraway's 'Gender' article, predate the inception of the term 'intersectionality' associated with Kimberlé Crenshaw and her 'Demarginalizing the Intersection of Race and Sex: A Black Feminist Critique of Antidiscrimination Doctrine, Feminist Theory and Antiracist Politics' *University of Chicago Legal Forum*, 1 (1989): 139–68.
48 Haraway, 'Gender', 145, emphasis original.

49 Hazel V. Carby, *Reconstructing Womanhood: The Emergence of the Afro-American Woman Novel* (New York: Oxford University Press, 1987) 25.
50 Haraway, 'Gender', 146.
51 Darlin' refers to the Woman as 'the animal lady'.
52 Wolfe with Elmer, 'Subject to Sacrifice' 101.
53 Jacques Derrida, *The Beast and the Sovereign Volume 1*, trans. Geoffrey Bennington (Chicago: Chicago University Press, 2009) 17–18.
54 Derrida, Beast *1*, 23.
55 Ibid., 9.
56 Ibid., 278.
57 Arguably this extraction of words of thanks is more humiliating in this film than her sexual assault, in which the withering returned gaze of the Woman either causes Cleek to simply give up on the assault or to lose his erection and be unable to continue.
58 In personal correspondence regarding the contrasting oral appetites of the Woman and Darlin', Ruth Lipschitz acutely commented, 'The "*cum panis*" of The Woman's outlaw orality is indeed a kinship of eating (heteropatriarchy's life) blood, and not one of sharing out (ginger)bread (men).' 16 April 2019.
59 Kristeva, *Powers*, 2.
60 Freud, 'Totem and Taboo', 187.
61 Kristeva, *Powers*, 57.
62 Ibid.
63 Ibid., 57–8, emphasis mine.
64 Ibid., emphasis mine.
65 Ibid., 61.
66 Kristeva affirmatively references Levi-Strauss with regard to the logical connection between the incest taboo and the exchange of discrete units. See *Powers*, 63. The major work referenced is Claude Levi-Strauss, *The Elementary Structures of Kinship*, eds. James Harle Bell and John Richard von Sturmer, Revised Edition (London: Eyre & Spottiswoode, 1969).
67 Claude Levi-Strauss, qtd. in Elizabeth Cowie, 'Woman as Sign' in *M/F*, 1 (1978): 52. Cowie's article was a key step in releasing feminist theory from a requirement for realism as the best method of accounting for the reality of women's experience and turning instead to an engagement with signification. In so doing, however, she maintained a logic of presence, given that sender, receivers and signs all retain their own integrity. For a deconstructive critique of the limits of this engagement see Lynn Turner '*Des Jeunes Nées*: For a Confusion of the Tongue, the Lip, & the Rim' in *Issues in Contemporary Culture & Aesthetics*, 1 (2005): 171–8.

68 Jacques Lacan, qtd. in Cowie 'Woman as Sign' 60, emphasis mine.
69 Kristeva, *Powers*, 58.
70 Ibid., 70.
71 See Judith Butler, 'Stubborn Attachment, Bodily Subjection: Rereading Hegel on the Unhappy Consciousness' in *The Psychic Life of Power* (Stanford University Press, 1997) 31–62.
72 Compare this wiliness, uncontrollability and absolute non-representability that Kristeva aligns with the maternal-feminine (cast in an animalized figure) with the force that Lee Edelman attempts to ascribe to the fundamentally non-reproductive and antisocial queer, he of 'No Future'. One might ponder Edelman's commitment to the Lacan of the 1950s in order to make this argument. See Lee Edelman, 'The Future Is Kid Stuff' in *No Future: Queer Theory & the Death Drive* (Durham & London: Duke University Press, 2004) 1–31.
73 See Alexandre Kojève, *Introduction to the Reading of Hegel: Lectures on the Phenomenology of Spirit*, [1947] trans. James H. Nichols, Jr. (Ithaca: Cornell University Press, 1980) 6–7.
74 Derrida, *Beast 1*, 102. I return to this problem in the final chapter.
75 Kristeva, *Powers*, 75.
76 Ibid., 70.
77 The term plays on the established overlap of speciesist and misogynist use of feminine animal nouns. See Carol Adams, *The Sexual Politics of Meat* [1990] (London: Bloomsbury, 2010).
78 The novel suggests that the Woman does identify with the dogs that she can hear through a carnivorous fantasy of wildness of 'tooth and claw,' 163
79 Kristeva, *Powers*, 72.
80 Ibid., emphasis added.
81 Ibid., 73.
82 Oliver, *Animal Lessons*, 293. Oliver perhaps over-reads Kristeva's amendment to Freud, i.e. it is somewhat misleading to claim that Kristeva departs from Freud in order to prioritise 'the mother over the father.'
83 See Derrida, *Animal*, 42–3.
84 Luce Irigaray did not simply deny Freud's identification of blindness with castration but derided his inability to see anything that is not the same as the phallus. See her canonical essay 'The Blindspot of an Old Dream of Symmetry' in *Speculum of the Other Woman*, trans. Gillian C. Gill (Ithaca: Cornell University Press, 1985): 11–132.
85 Sigmund Freud, qtd. in Ranjanna Khanna, *Dark Continents: Psychoanalysis and Colonialism* (Durham: Duke University Press, 2003) 47.

86 Wolfe with Elmer, 'Subject to Sacrifice', 101.
87 Carol Adams, 'Why Feminist-Vegan Now?' in *Feminism & Psychology*, 20:3 (2010): 303–17.
88 Derrida, *Beast 1*, 23.
89 See Derrida *Beast 1*, 65: 'And as for orality between mouth and maw, we have already seen its double carry [*portée*], the double tongue, the carry of the tongue that speaks, carry's the carry of the voice that *voci*ferates (to voci-ferate is to *carry the voice*) and the other carry, the other devouring one, the *voracious carry* of the maw and the teeth that lacerate and cut to pieces'.
90 At 5'11', the actress, Pollyanna McIntosh, casts an imposing figure.
91 See Jacques Derrida, 'Ninth Session' in *The Death Penalty V.II*, trans. Elizabeth Rottenberg (Chicago: Chicago University Press, 2017) 288–324.
92 Cary Wolfe, 'Learning from Temple Grandin: Animal Studies, Disability Studies and Who Comes After the Subject' in *What Is Posthumanism?* (Minneapolis: Minnesota University Press, 2010). Wolfe summarises it thus: 'it is the blind, the *dis*abled, who "see" the truth of vision. It's the blind who most readily understand that the core fantasy of humanism's trope of vision is to think that perceptual space is organized around and for the looking subject: that the pure point of the eye (as agent of *ratio* and *logos*) exhausts the field of the visible; that the "invisible" is only-indeed merely-that which has not yet been seen by a subject who is, in principle, capable of seeing all.' 132, emphasis original.
93 Oliver, *Animal Lessons*, 247–8.
94 Terri Tomsky acutely observes how prisoners at Guantánomo were severely punished for implicitly taking other animals (in this case iguanas) as pets by surreptitiously feeding them and in so doing gaining a feel of their own humanity. The gesture of domestication was seen to participate in the 'normative' construction of humanity construed according to 'ideas of power, privilege, belonging, self-assurance and self-definition' from which they were barred. Tomsky also notes that the regime at Guantánomo abjected the prisoners not just from the category of the 'human' but even from that of the 'animal'. Below both, the prisoners lobbied to have the same rights they saw accorded to animals (notably the guard dogs). See Terri Tomsky, 'Iguanas and Enemy Combatants: Reconsidering Cosmopolitanism through Guantánamo's Creaturely Lives' in Kaori Nagai, Karen Jones, Donna Landry et al. eds. *Cosmopolitan Animals* (Basingstoke: Palgrave, 2015) 205–8.
95 The last chapter of this volume develops Derrida's work on the death penalty and cruelty in which cruelty draws on 'cruor': blood that flows.

96 Oliver, *Animal Lessons*, 126.
97 This other ending is so secret that I missed it for some time! I thank a postgraduate student from my MA module, 'Sex Gender Species', in which this material was tested, for alerting me to it.

Chapter 4

1 Jacques Derrida, 'To Speculate – on "Freud"', in *The Post Card: From Socrates to Freud and Beyond*, trans. Alan Bass (Chicago: Chicago University Press, 1987) 344. Derrida is keeping Freud to his word, his 'often far-fetched speculation'.
2 Luce Irigaray, 'The Blindspot in an Old Dream of Symmetry' in *Speculum of the Other Woman* [1974], trans. Gillian C. Gill (Ithaca: Cornell University Press, 1985) 51. Reproduced with the permission of Cornell University Press.
3 Ada, Countess of Lovelace, qtd. in Betty Alexandra Toole, *Ada, The Enchantress of Numbers: A Selection from the Letters of Lord Byron's Daughter and Her Description of the First Computer* (California: Strawberry Press, 1992).
4 See, for example, Griselda Pollock, 'Killing Men & Dying Women: A Woman's Touch in the Cold Zone of American Painting in the 1950's' in Griselda Pollock and Fred Orton, eds. *Avant-Gardes & Partisans Reviewed* (Manchester: Manchester University Press, 1996) 221–94, or Hilary Robinson, *Reading Art, Reading Irigaray: The Politics of Art by Women* (London: I.B. Tauris, 2006).
5 See Sigmund Freud, 'Beyond the Pleasure Principle' [1920] in *On Metapsychology*, The Penguin Freud Library V. 11, trans. Strachey, James (London: Penguin, 1991) 283–86. The dialectical logic that drives man to overcome this initial privation is elaborated in Chapter 6 of this book in the context of the vanity of anthropocentrism.
6 Derrida, 'To Speculate', 268.
7 Luce Irigaray, 'Gesture in Psychoanalysis' [1985] in *Sexes & Genealogies*, trans. Gillian C. Gill (New York: Columbia University Press, 1989) 89–104.
8 Swinton became known as an actress in the experimental early queer films of Derek Jarman such as *The Last of England* (1987) Film} Dir. Derek Jarman (UK/West Germany:Anglo Internalional Films).
9 Derrida, 'Envois', 113, 159.
10 See Hershman Leeson's startlingly brilliant performance project which constructed the identity of one Roberta Breitmore, an identity that could be performed by herself or another and even extended to a social security number

and visits to a therapist. See Meredith Tromble, *The Art & Films of Lynn Hershman Leeson: Secret Agents, Private I* (Berkeley: University of California Press, 2005). See also her later documentary concerning the emergence of feminist art in the United States: *!Women Art Revolution! A (Formerly) Secret History* (2010) [Film] Dir. Lynn Hershman Leeson, USA: Hotwire Productions.

11 *Conceiving Ada* (1997), [Film] Dir. Lynn Hershman Leeson, USA: Hotwire Productions), trailer accessible here: https://www.imdb.com/title/tt0118882/?ref_=nv_sr_srsg_0. Accessed 20 May 2020.

12 Founded in 2009 by Suw Charman-Anderson, Ada Lovelace Day is now held every year on the second Tuesday of October. It features the flagship *Ada Lovelace Day Live!* 'science cabaret' in London, UK, at which women in STEM give short talks about their work or research in an informal, theatre-like setting. https://findingada.com/. Accessed 20 May 2020.

13 *Absent Presence* (2005), [Film] Dir. Hussein Chalayan, UK/Turkey: BM Contemporary Art Centre, 15 minutes) was also exhibited as a five-screen installation in the Turkish Pavilion at the Venice Biennale, 2005. The whole film can be seen here: https://vimeo.com/97742845. Accessed 20 May 2020. While they have not had the same level of exposure, Chalayan has made several other short films that also engage genetic anthropology such as *Temporal Meditations* (2003) and *The Art of Fashion* (2009).

14 These seminars are generating a great deal of attention with the French edition now published: Jacques Derrida, *La vie la mort séminaire (1975–1976)* (Paris: Seuil, 2019). For a significant commentary, see Dawne McCance, *The Reproduction of Life Death: Derrida's La vie la mort* (New York: Fordham University Press, 2019).

15 McCance, *Life Death*, 11.

16 Ibid., 11–12.

17 Freud, 'Beyond the Pleasure Principle', 334.

18 Ibid., 275.

19 Ibid., 283.

20 I developed this line – incorporating the SF of Sandor Ferenczi and the latter's anxiety regarding his own speculative departures from the House of Freud – in my 'Fort Spa: In at the Deep End with Derrida and Ferenczi' in Simon Morgan Wortham and Chiara Alfano, eds. *Desire in Ashes: Deconstruction, Psychoanalysis, Philosophy* (London: Bloomsbury, 2015) 103–20.

21 My 'Fort Spa' also noted, as will become the case here, that this signature, SF, is necessarily promiscuous. See Donna Haraway's Pilgrim Award acceptance

comments: 'SF: Science Fiction, Speculative Fabulation, String Figures, So Far' in *SFRA Review,* 297 (2011): 12–19.
22 Freud, 'Beyond the Pleasure Principle', 276, emphasis original.
23 Derrida, 'To Speculate', 299.
24 Ibid., 317. See Freud, 'Beyond the Pleasure Principle', 284.
25 Derrida, Jacques, 'Envois', in *The Post Card: From Socrates to Freud & Beyond*, trans. Alan Bass (Chicago: Chicago University Press, 1987) 66, emphasis added.
26 Freud, 'Beyond the Pleasure Principle', 285.
27 Ibid., 275.
28 Derrida, 'To Speculate', 306.
29 Ibid., 308.
30 Freud, 'Beyond the Pleasure Principle', 284.
31 Derrida, 'To Speculate', 314.
32 Ibid., 315.
33 Ibid.
34 Derrida 'To Speculate', 322. At this point Derrida exchanges '*fort/da*' with '*fort:da*'.
35 Ibid., 283.
36 Ibid., 318.
37 Ibid. See Freud, 'Beyond the Pleasure Principle', 284, n.1.
38 See the discussion of Derrida's critical relationship to the mirror of autobiography in Chapter 6 of this volume.
39 Derrida, 'To Speculate', 320.
40 Ibid., 322.
41 Ibid.
42 Freud, 'Beyond the Pleasure Principle', 285.
43 I.e. Go to the war (and don't come back). Ibid., 286.
44 Ibid., 285, n.1.
45 Derrida, 'To Speculate', 283.
46 Ibid., 269.
47 Ibid., 269.
48 Ibid., 283.
49 See Derrida, 'To Speculate', 342–3.
50 Ibid., 285.
51 Ibid., 285.
52 Ibid., 293
53 Ibid., 339.

54 Foreshadowed in the opening credits when the letters *o* and *i* of 'Conceiving' are animated and effectively mate. Is this how zeroes and ones reproduce?
55 Sharon Lin Tay, '*Conceiving Ada*: Conceiving Feminist Possibilities in the New Mediascape' in *Women: A Cultural Review*, 18:2 (2007): 188. One could counter all of Tay's assumptions. For example, she suggests that *Conceiving Ada* resists the formula of the classic realist film text (beginning with a disequilibrium and ending with restoration), yet the film precisely carries this out as a flawed ethical imperative (the disequilibrium is the lack of justice to women, especially Ada: its restoration is the revivification of Ada in the child born in our present time of emancipation).
56 Carole Cadwalladr is a key journalist reporting on the still ongoing scandal of Facebook's harvesting and management of data and its relationship with amorphous companies such as Cambridge Analytica. See her TED talk: https://www.ted.com/speakers/carole_cadwalladr
57 Ada Lovelace as A. A. L. translated and added substantial notes to Federico Luigi, Conte Menabrea's 'Sketch of the Analytical Engine Invented by Charles Babbage, Esq' (1842) in *Scientific Memoirs*, V.3. See Betty Alexandra, Toole, *Ada, the Enchantress of Numbers: A Selection from the Letters of Lord Byron's Daughter & Her Description of the First Computer* (Mill Valley, CA: Strawberry Press, 1992).
58 According to Doron Swade, 'Ada saw something that Babbage in some sense failed to see. In Babbage's world his engines were bound by number… What Lovelace saw – what Ada Byron saw – was that number could represent entities other than quantity. So once you had a machine for manipulating numbers, if those numbers represented other things, letters, musical notes, then the machine could manipulate symbols of which number was one instance, according to rules. It is this fundamental transition from a machine which is a number cruncher to a machine for manipulating symbols according to rules that is the fundamental transition from calculation to computation – to general-purpose computation – and looking back from the present high ground of modern computing, if we are looking and sifting history for that transition, then that transition was made explicitly by Ada.' See Fuegi, J; Francis, J, 'Lovelace & Babbage and the Creation of the 1843 "Notes"' in *Annals of the History of Computing*, 25:4 (October–December 2003): 16–26.
59 I am paraphrasing the film's script. For an account of Ada's gambling, see Benjamin Woolley, *The Bride of Science: Romance, Reason, and Byron's Daughter* (Basingstoke: Pan Macmillan, 1999).
60 Irigaray, 'Gesture', 95.
61 Irigaray has written of this expulsion, along with that from her teaching post at the University of Paris, Vincennes, in her book co-authored with Michael

Marder, *Through Vegetal Being* (New York: Columbia University Press, 2016) 14–16. She does not name Lacan.

62 Amy M. Hollywood, 'Deconstructing Belief: Irigaray & the Philosophy of Religion' in *The Journal of Religion*, 78:2 (1998): 236, n.17. It is unfortunate that no sustained conversation ever emerged between Derrida and Irigaray. The frankly rather strange handful of footnotes indexing Irigaray in *On Touching-Jean-Luc Nancy* carefully date their references to publications from *This Sex Which Is Not One* onwards (listing only English translations), thus bypassing *Speculum*. One note even remarks that Derrida is 'not certain' whether Irigaray's work 'intersects' with that of Nancy or not. See Derrida *On Touching-Jean-Luc Nancy* [2000] trans. Christine Irizarry (Stanford: Stanford University Press, 2005) 332, n.31.

63 Luce Irigaray, 'Belief Itself' [1980] in *Sexes & Genealogies*, trans. Gill, Gillian C. (New York: Columbia University Press, 1993) 23–53. This elliptical poetic text has attracted substantially less commentary than has 'Gesture'. Hollywood's work is a notable exception; see 'Deconstructing Belief'.

64 Irigaray, 'Gesture', 97. She does not mention the conference was on Derrida's work.

65 Ibid.

66 Irigaray, 'Blind Spot', 78, emphasis original.

67 Irigaray, 'Belief Itself', 25.

68 Jacques Derrida, 'Le Facteur de la Verité' [1975] in *The Post Card: From Socrates to Freud and Beyond*, trans. Alan Bass (Chicago: Chicago University Press, 1987) 411–98.

69 Derrida, 'Facteur de la Verité', 444, emphasis added. The infamous 'exchange of letters' has produced its own archive, key texts of which are to be found here: John P. Miller and William J. Richardson, eds. *The Purloined Poe: Lacan, Derrida and Psychoanalytic Reading* (Baltimore: Johns Hopkins University Press, 1987).

70 Derrida, 'Facteur de la Verité', 442.

71 Ibid., 444.

72 Irigaray, 'Belief Itself', 25. Luce Irigaray, *This Sex Which Is Not One* [1977] trans. Catherine Porter (Ithaca: Cornell University Press, 1985) 78. (Irigaray's discontinuity with Kristeva can be gleaned from this gesture.)

73 Irigaray, 'Gesture', 92.

74 Ibid.

75 Ibid. 93–4.

76 Irigaray, 'Gesture', 92. Frankly this could be taken at face value as the persistent condition of everyday sexism, but in this context of a genealogical economy, it sounds resentful.
77 Ibid.
78 See Penelope Deutscher's investigation of 'eating the other' between Irigaray and Derrida in 'Mourning the Other, Cultural Cannibalism and the Politics of Friendship (Jacques Derrida and Luce Irigaray)' in *Differences: A Journal of Feminist Cultural Studies*, 10:3 (1998): 159–84.
79 Ibid. 98.
80 See Luce Irigaray, 'When Our Lips Speak Together' in *This Sex Which Is Not One*, trans. Catherine Porter (Ithaca: Cornell University Press, 1985) 205–18.
81 Irigaray, 'Gesture', 98.
82 Ibid. 95–6.
83 Ibid. 97.
84 Ibid.
85 Irigaray, 'Belief Itself', 31.
86 Irigaray, 'Gesture', 97.
87 Irigaray, 'Belief Itself', 31.
88 Irigaray's 'Gesture' hyphenates the game – *fort-da* – whereas Derrida uses a forward slash until he makes the point of the possession of da when the reel is shown to never really go away – *fort:da*.
89 Irigaray, 'Gesture', 99. Arguably this both accepts a discrete choreography of the masturbating boy (entirely penile, relentlessly outwardly directional) and accepts a discrete map of the working of his body. Compare the speculative 'genital amphimixis' proposed by Sandor Ferenczi discussed in my 'Fort Spa: In at the Deep End with Derrida and Ferenczi', in *Desire in Ashes*.
90 Irigaray, 'Gesture', 99
91 Ibid.
92 Derrida, 'To Speculate', 292.
93 Irigaray, 'Gesture', 100, emphasis original.
94 Ibid. Incidentally, the third section of 'To Speculate' is named 'Paralysis'.
95 Irigaray, 'Belief Itself', 28.
96 Derrida, 'To Speculate', 306.
97 For her derisive citation of Freud's account of the little girl's genitals as posing 'nothing to be seen', see Irigaray, 'Blind Spot', 47–8.
98 Irigaray, 'Belief Itself', 38, n.6.
99 Derrida, 'To Speculate', 341.

100 The accusation of 'essentialism' has severely hindered Irigaray's Anglophone reception, partly due to the dissonance between 'gender' and 'sexual difference'.
101 Ada used various bird names for herself and others, including 'Carrier Pigeon'. See Toole, 30, for example.
102 Johnny Golding heard these homonyms when listening to early drafts of this chapter. In light of the question that Derrida often asks himself, discussed in Chapter 6 in this book, we could also ask here, 'Who am I?' and thus 'Who am I following?'
103 Irigaray, 'Belief Itself', 41.
104 The bird contrasts Emmy's other, earlier, agent the virtual copy of her dog she names Gods-Dog. As a virtual interlocutor, Gods-Dog tells tales on Emmy, preventing her from keeping a secret.
105 See Freud, 'Beyond the Pleasure Principle', 312.
106 Derrida, 'To Speculate-on "Freud"', 358.
107 Dawne McCance, 'Death' in Lynn Turner, Undine Sellbach and Ron Broglio, eds. *The Edinburgh Companion to Animal Studies* (Edinburgh: Edinburgh University Press, 2018).
108 McCance, 'Death', 121.
109 Compare Swinton in her titular role in *Orlando* (dir. Sally Potter, UK, 1992).
110 Not only does palindrome indicate a proper name that can be spelled backwards and forwards, it also indicates a segment of DNA in which the nucleotide sequence in one strand read from end is the same as the sequence in the complementary strand read from the opposite end.
111 Mignon Nixon, 'The She-Fox: Transference & the "Woman Artist"', in Catherine Armstrong and Catherine de Zegher, eds. *Women Artists at the Millennium* (Cambridge: MIT Press, 2006) 275–301.
112 Sigmund Freud, 'Some Reflections on Schoolboy Psychology' [1914] cited in Nixon, 'She-Fox', 277–8.
113 Nixon, 'She-Fox', 278. Nixon also notes the path that Melanie Klein opened for examining precisely negative transference in women.
114 Ibid., 277.
115 Freud, qtd. in Nixon, 'She-Fox', 276.
116 There is some biographical support for the heavy-handed control of Lady Byron, desperate to offset any inherited wildness that Ada might inherit from her Romantic father. See James Essinger, *Ada's Algorithm: How Lord Byron's Daughter Launched the Digital Age through the Poetry of Numbers* (London: Gibson Square, 2017) 46–8. However the feminist point remains: why concentrate on the mother as the root of Ada's woes?

117 Derrida, 'To Speculate', 356.
118 Ibid., 355.
119 Jacques Derrida, *Voice and Phenomenon* [1967] trans. Leonard Lawlor (Evanston: Northwestern University Press, 2011) 65, emphasis added. The citation within this quotation is from Husserl, whose transcendental reduction is here under Derrida's review.
120 McCance, *Life Death*, 71. McCance reports that the nineteenth-century eugenics movement explicitly linked 'deafmutes' with 'foreigners' and with immigration (92).
121 Imagining the computer as 'wetware' is a 'dry' inversion of the opposition circulated in early cyberpunk novels in which the 'hardware' of technology was counterposed by the 'wetware' of the body.
122 See Paul de Man, 'The Concept of Irony' in *Aesthetic Ideology* (Minneapolis: Minnesota University Press, 1996) 165-6.
123 Jacques Derrida, 'Plato's Pharmacy' in *Dissemination* [1972] trans. Barbara Johnson (London: Athlone, 1981) 63.
124 McCance, *Life Death*, 139.

Chapter 5

1 Hélène Cixous, 'Writing Blind: Conversation with the Donkey' [1996], trans. Eric Prenowitz, ed. in *Stigmata: Escaping Texts* (New York and London: Routledge, 1997) 140. Reprinted by permission of the press.
2 The first two papers published in English translation from Derrida's address at the 1997 colloquium, *L'Animal Autobiographique*, were: Derrida, Jacques. 'The Animal That Therefore I Am (More to Follow)', trans. David Wills. *Critical Inquiry* 28:2 (2002): 369-418 and 'And Say the Animal Responded?' trans. David Wills, in Cary Wolfe, ed. *Zoontologies: The Question of the Animal* (Minneapolis: Minnesota University Press, 2003) 121-46.
3 See Lynn Turner, 'When Species Kiss: Some Recent Correspondence between *Animots*' in *Humanimalia: A Journal of Human/Animal Interface Studies*, 2:1 (2010): 60-85. www.depauw.edu/humanimalia/. A shorter piece was also published that year, bringing Derrida's unconditional ethics and Haraway's companion species into proximity with Schneemann's *Infinity Kisses*: Carla Benzan, 'The Lives and Deaths of Carolee's Cats: Intimate Encounters, Gentle Transgressions and Incalculable Ethics' in *C Magazine International Contemporary Art*, 107 (Autumn 2010): 5-11. Benzan also remarked on the

paucity of scholarship that would admit the feline subject matter into the feminist canon of Schneemann's practice.

4. Donna J. Haraway, *The Companion Species Manifesto: Dogs, People & Significant Otherness* (Chicago: Prickly Paradigm Press, 2003); Donna J. Haraway, *When Species Meet* (Minneapolis: Minnesota University Press, 2008); Hélène Cixous, 'The Cat's Arrival' in *Parallax,* 12:1 (2006): 22. While the metallic gleam of the 'cyborg' seems to still distract some readers – Haraway's announced 'irony' notwithstanding – her development of relating with non-human species should not have come as a surprise from the author who once wrote, at length, of 'primatology' as a 'genre of feminist theory': *Primate Visions: Gender, Race, and Nature in the World of Modern Science* (New York: Routledge, 1989) 279–382.

5. With regard to Cixous and Haraway, in particular, this chapter builds on my subsequent publications, especially: Lynn Turner, ed. *The Animal Question in Deconstruction* (Edinburgh: Edinburgh University Press, 2013); 'Critical Companions: Derrida, Haraway & Other Animals' in Julian Wolfreys, ed. *Introducing Criticism in the 21st Century*, 2nd Edition (Edinburgh: Edinburgh University Press, 2015) 63–80; 'Telefoam: Species on the Shores of Cixous and Derrida' in Ivan Callus, Stefan Herbrechter and Manuela Rossini, eds. *European Posthumanism* (London and New York: Routledge, 2016) 56–69.

6. The foundational text here remains Jacques Derrida, 'Signature Event Context' in *Margins of Philosophy*, trans. Alan Bass (Chicago: University of Chicago Press, 1982). Infamously, 'SEC' was reproduced along with further extended lessons on the vicissitudes of performative speech acts in Derrida's *Limited Inc* (Evanston: Northwestern University Press, 1988).

7. The popular uptake of 'the performative' largely drew on the work of Judith Butler and Peggy Phelan, e.g. *Gender Trouble: Feminism and the Subversion of Identity* (New York and London: Routledge, 1990) and *Unmarked: The Politics of Performance* (New York and London: Routledge, 1993).

8. Jacques Derrida, 'Composing "Circumfession"' in Caputo, John D. and Michael J. Scanlon, eds. *Augustine & Postmodernism: Confessions & Circumfession* (Bloomington: Indiana University Press, 2005) 21.

9. See my 'Wind Up: The Machine-Event of Tape' in Astrid Schmetterling and Lynn Turner, *Visual Cultures as … Recollection* (Berlin: Sternberg, 2013) 29–52.

10. See Jacques Derrida, 'Typewriter Ribbon: Limited Ink (2)' in *Without Alibi*, trans. Peggy Kamuf (Stanford: Stanford University Press, 2002) 71–160 (first given as a keynote address at the *Material Events: Paul de Man and the AfterLife of Theory* conference at the University of California, 2000).

11 Derrida, 'Typewriter Ribbon', 73.
12 In my 'Insect Asides' I followed the insects preserved in amber, briefly and elliptically referenced in 'Typewriter Ribbon', in Turner, ed. *The Animal Question in Deconstruction*, 54–69.
13 Derrida, 'Typewriter Ribbon', 74.
14 See René Descartes, *The Philosophical Writings of René Descartes*, V.1, trans. John Cottingham, Robert Stoothoff and Dugald Murdoc (Cambridge: Cambridge University Press, 1985) 108.
15 Karen Barad, 'Posthumanist Performativity: Toward an Understanding of How Matter Comes to Matter' in *Signs*, 28:3 (2003): 801–31. Haraway's *When Species Meet* was the third book to be published in what is now a highly influential book series: *Posthumanities*, edited by Cary Wolfe.
16 On this distinction see Ivan Callus, Stefan Herbrechter and Manuela Rossini, 'Dis/locating Posthumanism in European and Literary Critical Traditions' in *European Journal of English Studies*, 18:2 (2014): 13–20.
17 Donna J. Haraway, *Staying with the Trouble: Making Kin in the Chthulucene* (Durham, NC: Duke University Press, 2016) 97. See also her 'Companions in Conversation' with Cary Wolfe in Donna J. Haraway, ed. *Manifestly Haraway* (Minneapolis: Minnesota University Press, 2016) 261.
18 Lynn Turner, Undine Sellbach, Ron Broglio, 'Introducing The Edinburgh Companion to Animal Studies' in Lynn Turner, Undine Sellbach and Ron Broglio, eds. *The Edinburgh Companion to Animal Studies* (Edinburgh: Edinburgh University Press, 2018) 8.
19 Turner, Sellbach, Broglio, eds. *Edinburgh Companion to Animal Studies*, 9.
20 Derrida, *Animal*, 51.
21 Ibid., 69–70. Descartes sentence reads: 'But as for me, whom am I [*qui suis-je*], now that I am supposing that there is some supremely powerful and, if it is permissible to say so, malicious deceiver, who is deliberately trying to trick me in every way he can?'
22 Derrida lists the 'malicious deceiver' in Descartes as 'the most cunning of the animals', *Animal*, 70.
23 Directly addressed in, for example, Cixous's *Portrait of Jacques Derrida as a Young Jewish Saint* (New York: Columbia University Press, 2004), and Derrida's *H.C for Life, That Is to Say …*, trans. Laurent Milesi (Palo Alto: Stanford University Press, 2006).
24 See Turner, 'Telefoam'.
25 Jacques Derrida, 'The Law of Genre' in Derek Attridge, ed. *Acts of Literature* (New York and London: Routledge, 1992) 221–52.

26 Hélène Cixous, 'From My Menagerie to Philosophy' in Dorothea Olkowski, ed. *Resistance, Flight, Creation: Feminist Enactments of French Philosophy* (Ithaca: Cornell University Press, 2000) 40, Hélène Cixous, *La* (Paris: Editions Gallimard, 1976) 91. Steve Baker first mentioned Cixous's early use of the term *animot* to me.

27 Cixous, 'Menagerie', 44. The English translation of 'Menagerie' appears with an epigraph from Derrida's text *'Che cos'è la poesia?'* (1988), in which poetry is likened to a hedgehog. Cixous's text also appears – without that epigraph, as 'De La Ménagerie à la Philosophie', the chapter succeeding 'Arrivée du Chat' in her novel *Messie* (Paris: Des Femmes, 1996).

28 *Interior Scroll*, first performed in 1975, subsequently performed on numerous occasions in various venues, sometimes incorporating group elements sometimes remaining the individual performance of Schneemann herself. *Interior Scroll* is also widely exhibited as photographic documentation – in vertical series that suggest both a strip of film and a scroll – including the text that Schneemann recited.

29 Thyrza Nichols Goodeve notes that these references were to the art historian Annette Michelson; see her '"The Cat Is My Medium:" Notes on the Writing and Art of Carolee Schneemann' *Art Journal Open*, 29 July 2015. http://artjournal.collegeart.org/?p=6381

30 This implicit 'ban' is discussed and revoked in Amelia Jones, *Body Art/Performing the Subject* (Minneapolis: University of Minnesota Press, 1998).

31 On the revised law of 'all "experience" in general', see Derrida, 'Signature Event Context', 318.

32 Haraway, *Companion Species Manifesto*, 2, italics original.

33 Cixous, 'Cat's Arrival', 22.

34 'Kiss Me Honey Honey Kiss Me' was a popular song written by Albon Timothy and Michael Julien, first recorded by Shirley Bassey in 1959.

35 http://www.caroleeschneemann.com/index.html

36 Carolee Schneeman, *Imaging Her Erotics: Essays, Interviews, Projects* (Cambridge, MA: MIT Press, 2002) 264.

37 *Eye/Body*, 1963; *Meat Joy*, 1964; *Interior Scroll*, 1975; *Up to and Including Her Limits*, 1976. First dates given: all performances performed on numerous occasions in various venues, sometimes incorporating group elements sometimes remaining the individual performance of the artist. *Fuses*, 18 min, colour, silent 16mm film, 1965. *Meat Joy* now warrants a whole other engagement with the use of animal meat as metaphor.

38 Jones, *Body Art*. It should be noted that Jones takes little account of the development of Pollock's extensive work beyond 1987.

39 Conversation with the artist, March 2008.
40 *Breaking Borders*, Museum of Contemporary Canadian Art, Toronto, 2007 and *Remains to Be Seen*, CEPA Gallery (Buffalo: New York, 2007). The curators were David Liss and Photios Giovanis.
41 See, for example, 'A Tribute to Carolee Schneemann (1939–2019)' *The Brooklyn Rail*, April 2019. Of the several contributors to this 'In Memoriam', Ann McCoy, Heide Hatry and Thryza Nichols Goodeve all acknowledged the radicality of the feline content of Schneemann's work. https://brooklynrail.org/2019/04/in-memoriam/A-Tribute-to-Carolee-Schneemann-1939-2019
42 Rebecca Schneider, *The Explicit Body in Performance* (New York: Routledge, 1997) 49.
43 Jacques Derrida, 'Choreographies'. Interview with Christie McDonald. *Points. Interviews, 1974–1994* (Palo Alto: Stanford University Press, 1994) 50.
44 Schneeman, *Imaging Her Erotics* 264.
45 Steve Baker, *The Postmodern Animal* (London: Reaktion, 2000) 170.
46 This question is asked, and answered in variously sentimental and sanctimoniously anti-anthropomorphic ways with regard to dogs on this website: http://ask.metafilter.com/28246/Is-my-dog-kissing-me. Accessed 23 November 2007.
47 J. Hillis Miller, 'What Is a kiss? Isabel's Moments of Decision' in *Critical Inquiry*, 31 (Spring 2005): 724.
48 Luce Irigaray works metaphor and metonymy together to address both a positive symbolization of the feminine through labial multiplicity (in excess of, rather than in opposition to, phallic unicity and the logic of the same) and to link this symbolization to a new political, poetic and symbolic speech and 'speech' by women most notably. See her 'When Our Lips Speak Together'.
49 Hillis Miller, 'What Is a kiss?', 731.
50 Derrida regularly distances the semiotically inclined and delimited 'polysemy' from dissemination as that which 'cannot be pinned down at any one *point*'. See Jacques Derrida, *Dissemination*, trans. Barbara Johnson (London: The Athlone Press, 1981) 58 italics original. See also Derrida, 'Heidegger's Hand', 57.
51 Hillis Miller, 'What Is a kiss?', 725.
52 Ibid., 728.
53 For an extended discussion of cetacean communication, see Lynn Turner, 'Voice' in Turner, Sellbach, Broglio, eds. *Edinburgh Companion to Animal Studies* 518–32.
54 Schneeman, *Imaging Her Erotics*, 264

55 Schneemann qtd. in Steve Baker, 'What Does Becoming-Animal Look Like?' in Nigel Rothfels, ed. *Representing Animals* (Bloomington: Indiana University Press, 2002) 73. Baker's subsequent and more sympathetic account of Schneemann doesn't remark on the technological tropes that facilitate Schneemann's remark (in favour of a more art historical reportage of the artist's account). I address this in the last section of this chapter.

56 Carol Adams, 'Preface to the Twentieth Anniversary Edition', in *The Sexual Politics of Meat* (London: Continuum, 2010).

57 Leonard Lawlor, *This Is Not Sufficient: An Essay on Animality and Human Nature in Derrida* (New York: Columbia University Press, 2007).

58 See Haraway, *When Species Meet*, 19–27.

59 Derrida frequently remarks on this accrual of credit by the being who calls himself 'Man'; see, for example, *Animal*, 30.

60 Derrida, *Animal*, 13.

61 Dawne McCance, 'Death' in Turner, Sellbach and Broglio, eds. *Edinburgh Companion to Animal Studies*, 117–19.

62 Jacques Derrida, '*Geschlecht* II: Heidegger's Hand' [1987], trans. John P. Leavey, Jr., *Psyche: Inventions of the Other*, V. II, eds. Peggy Kamuf and Elizabeth Rottenberg (Stanford University Press, 2008) 42.

63 Heidegger, qtd. in Derrida, 'Heidegger's Hand', 43.

64 As I first advocated in my 'Critical Companions'.

65 Derrida, *Animal*, 34.

66 Ibid., 62–4, emphasis added. See Turner, 'Critical Companions'.

67 Derrida, qtd. in Lawlor, *This Is Not Sufficient*, 50.

68 Heidegger, qtd. Lawlor, *This Is Not Sufficient*, 51. See also Derrida, *Animal*, 36. Heidegger's three theses on 'world', from his *The Fundamental Problems of Metaphysics*, have now drawn myriad commentaries in animal studies.

69 Lawlor, *This Is Not Sufficient*, 51.

70 Ibid., 53.

71 Lawlor, *This Is Not Sufficient*, 55.

72 Derrida, *Animal*, 26. Derrida gives greater indication of his critical relation to the discourse of 'rights' in the interview 'Violence Against Animals', in Derrida and Elizabeth Roudinesco, eds. *For What Tomorrow ...*, trans. Jeff Fort (Stanford: Stanford University Press, 2004) 62–76. This should not be taken for any kind of disengagement with the plight of actual animals – whose deadly instrumentalization at the hands of Man is absolutely under scrutiny here. For a detailed account of the problems within philosophies of animal rights in light of

the ethics of deconstruction, see Dawne McCance, *Critical Animal Studies: An Introduction* (New York: SUNY, 2013).
73 See Lawlor, *This Is Not Sufficient*, 77–9.
74 Derrida, *Animal*, 14. Derrida too does not cite ethological sources but addresses himself to the pronouncements that philosophy has made regarding the animal.
75 Haraway, *When Species Meet*, 21.
76 Ibid., 24.
77 Haraway, *Companion Species*, 2.
78 Ibid., 1.
79 Ibid. Manuela Rossini is one of few scholars to engage the erotic play that is also very much at stake in this text. See her 'Coming Together: Symbiogenesis and Metamorphosis in Paul di Filippo's A Mouthful of Tongues' in Tom Tyler and Manuela Rossini, eds. *Animal Encounters* (Leiden: Brill, 2009) 243–58.
80 That Derrida writes that it is the 'ear of the other that signs' is hardly an otocentrism (especially if one eats well with 'all the senses in general'). See Jacques Derrida, 'Roundtable on Autobiography', trans. Peggy Kamuf in *The Ear of the Other: Otobiography, Transference, Translation* (New York: Shocken Books, 1985) 51.
81 Thryza Nichols Goodeve 'Nothing Comes Without Its World: Donna J. Haraway in Conversation with Thryza Nichols Goodeve – 20th Anniversary of *Modest_Witness*' in *Modest_Witness@Second_Millenium.FemaleMan©_Meets_Oncomouse™*(New York and London: Routledge, 2018) xiiiv.
82 Donna J. Haraway, '"Gender" for a Marxist Dictionary' [1987] in *Simians, Cyborgs & Women: The Reinvention of Nature* (London: Free Association Books, 1991).
83 Haraway, *Companion Species*, 2.
84 Haraway, *When Species Meet*, 11.
85 Ibid.
86 Derrida, *Animal*, 27; Haraway, *When Species Meet*, 22.
87 Derrida, *Animal*, 27.
88 Ibid.
89 Haraway, *When Species Meet*, 240.
90 See Jacques Derrida, 'Structure, Sign and Play in the Discourse of the Human Sciences' in *Writing and Difference*, trans. Alan Bass (London: Routledge and Kegan Paul, 1978) 289.

91 Along with Haraway's sometimes frustrated, sometimes hospitable relationship to Derrida (elaborated in my 'Critical Companions'), she has also been infuriated by what she perceived as an anti-domestic romance of the wild in the work of Gilles Deleuze and Felix Guattari. See Haraway, *When Species Meet*, 27–30. Ronald Bogue gives perhaps the most detailed response to this in his 'The Companion Cyborg: Technics and Domestication' in Jon Roffe and Hannah Stark, eds. *Deleuze and the Non/Human* (Basingstoke: Palgrave, 2015) 163–79.

92 Haraway, *Companion Species*, 2. Since Haraway's *Manifesto* was published, there has been a massive increase in theorization of microbial life; see, for example, Astrid Schrader, 'Microbial Suicide: Towards a Less Anthropocentric Ontology of Life and Death' in *Body and Society*, 23:3 (2017): 48–74.

93 Haraway, *Companion Species*, 103.

94 See Butler, *Gender Trouble* and in particular her subsequent *Antigone's Claim: Kinship between Life and Death* (New York: Columbia University Press, 2000).

95 Marc Shell, 'The Family Pet' in *Representations*, 15 (Summer 1986) 126.

96 Karen Joy Fowler's novel *We Are All Completely beside Ourselves* (New York: Serpent's Tail, 2013) is situated in the aftermath of a human family's adoption of a chimpanzee. As with historical cases such as that of Nim Chimpsky, the aim was to teach the chimp sign-language and include the chimp as one of the human family. The novel suggests that the learning relations are multi-directional and the human daughter has been as much shaped by this experience as has her chimp sibling.

97 Shell, 'Family Pet', 130. (He is referring to the Gospel of Paul.)

98 Ibid.

99 Shell, 'Family Pet', 141.

100 Derrida, 'Signature Event Context', 327. Note also Derrida's critique of the Freud who purported to only *describe* 'femininity': Jacques Derrida, 'Le Facteur de la Vérité', in *The Post Card: From Socrates to Freud & Beyond*, trans. Alan Bass (Chicago: Chicago University Press, 1987) 481, n. 60.

101 Donna J. Haraway, Modest_Witness@Second_Millenium.FemaleMan©_Meets_Oncomouse˜ (New York and London: Routledge, 1997) 265.

102 Haraway, *Companion Species*, 11.

103 Derrida, *Animal*, 12.

104 Haraway, *Companion Species*, 96.

105 Donna J. Haraway, *Staying with the Trouble: Making Kin in the Chthulucene* (Durham: Duke University Press, 2016) 103.

106 Haraway, *Staying with the Trouble*, 6.

107 As anticipated, Haraway drew deeply allergic responses to 'Make kin, not babies!', including from those she 'holds dear as "our people" on the Left' (*Staying with the Trouble*, 208, n.18). One profoundly misguided review of the book directly accused Haraway and Anna Tsing (!) of genocidal fantasy: 'Haraway's former (profoundly system-oriented) Marxian technofeminism has given way, then, to something called multispecies feminism: a tendency pioneered also by Anna Tsing characterized by a barely disavowed willingness to see whole cities and cultures wiped from the planet for the sake of a form of thriving among "companion species" involving relatively few of us.' See Sophie Lewis, 'Cthulhu Plays No Role for Me' *Viewpoint Magazine*, 8 May 2017. https://www.viewpointmag.com/2017/05/08/cthulhu-plays-no-role-for-me/
108 Haraway, *Staying with the Trouble*, 140.
109 Ibid., 139. Again, Haraway is careful to set up groups who choose symbionts *and* those who reproduce by choice without so doing, in explicit avoidance of eugenic selection.
110 Derrida, *Animal*, 28.
111 Shell, 'Family Pet', 141.
112 Ibid., italics original.
113 Haraway, *When Species Meet*, 22–3.
114 Her Catholic upbringing is referenced several times in this interview: Donna J. Haraway, *How Like a Leaf: An Interview with Thryza Nichols Goodeve* (London and New York: Routledge, 2000) 24.
115 Derrida, *Animal*, 3–4, italics original.
116 Ibid., 4, emphasis original.
117 Ibid., 17, emphasis added.
118 Ibid., 4.
119 Ibid., emphasis added.
120 Ibid., 50. Compare the *destinerrance* of the letter – even the self-addressed letter – of the postal principle discussed in Chapter 4.
121 Derrida, *Animal*, 1, emphasis added.
122 See Turner, 'Critical Companions'.
123 Derrida, *Animal*, 5.
124 Ibid.
125 Haraway, *Modest_Witness*, 23.
126 Ibid. 24.
127 Ibid. While Haraway – together with other feminist science studies scholars – insists that the experimental life was key to gender 'in the making', and not least

since women were excluded lest their empathetic engagement should interrupt proceedings, she later remarks that she was surprised at her own silence on the conscripted labour of the animals involved (e.g. the bird asphyxiating in the air pump). See Haraway, 'Nothing Comes Without Its World', xiii–xviii.

128 For a discussion of the overlapping relations between Haraway's reworking of the 'modest witness' and her partial or 'situated' knowledge, see 'Nothing Comes Without Its World'.

129 Steven Shapin and Simon Schaffer, *Leviathan and the Air Pump: Hobbes, Boyle and the Experimental Life* (Princeton: Princeton University Press, 1985) 66, emphasis original.

130 Derrida, *Animal*, 4.

131 Ibid., 50.

132 Jacques Derrida, '*Psyché*: The Invention of the Other' in Peggy Kamuf and Elizabeth Rottenberg, eds. *Psyché: The Invention of the Other V. 1* (Stanford University Press, 2007) 1–47. The force of this 'invention' turns against the calculation of the other.

133 Derrida asks after Lacan's conservatism regarding 'the animal' in 'And Say the Animal Responded'. While he discusses a number of essays in Lacan's *Ecrits*, the most famous and influential one at stake in this reflection is 'The Mirror Stage as Formative of the *I* Function as Revealed in Psychoanalytic Experience' [1949] in *Ecrits*, trans. Bruce Fink (New York: W. W. Norton, 2005) 75–81. The 'mirror recognition test' was developed in 1970 by psychologist Gordon Gallup, Jr. to attempt to establish whether non-human animals could recognize themselves. Notwithstanding the anthromorphic yardstick of what it is to recognize one's kind, some species, including the Great Apes, Cetaceans and Corvids, have been able to do so. Michael Zizer suggests that Lacan was quite aware of ethology contemporary to his own writing, and that would have challenged his specular system if admitted; see 'Animal Mirrors' in *Angelaki*, 12:3 (2007): 11–33.

134 While one might argue that the mirror stage does not offer a point-for-point reflection since the immature infant misrecognizes itself as more capable in the mirror, that misrecognition is teleologically endowed to impel the emergence of human subjectivity.

135 Cixous, 'Cat's Arrival', 21.

136 Ibid.

137 Marta Segarra, ed. *The Portable Cixous* (New York: Columbia University Press, 2010).

138 Turner, ed. *The Animal Question in Deconstruction*.

139 Hélène Cixous, 'Writing Blind: Conversation with the Donkey' [1996], trans. Eric Prenowitz, in *Stigmata: Escaping Texts* (New York and London: Routledge, 1997) 144. See also Turner, 'Telefoam', for further commentary on the Anglophone reception of Cixous.
140 Cixous, 'Cat's Arrival', 16.
141 David Wood named Derrida's insistence of the overlap of the domestic sphere and philosophical process as 'uncanny'. See his 'Thinking with Cats' in Matthew Calarco and Peter Atterton, eds. *Animal Philosophy: Essential Readings in Continental Thought* (London: Continuum, 2004) 132.
142 Derrida, 'Typewriter Ribbon', 73.
143 Jacques Derrida, *Aporias*, trans. Thomas Dutoit (Palo Alto: Stanford University Press, 1993) 86, n.14.
144 For a useful – though highly modest – discussion of these overlapping themes in their work, see Hélène Cixous, 'Jacques Derrida as a Proteus Unbound', trans. Peggy Kamuf, in *Critical Inquiry*, 33 (2007): 389–423.
145 Derrida, *Aporias*, 35.
146 Hélène Cixous and Catherine Clément, *La Jeune Née* (Paris: Union Générales d'Editions, 1975).
147 Cixous, 'Cat's Arrival', 21.
148 See Turner, 'Telefoam', for a discussion of the telephone in Cixous (prompted by Derrida's remark: '[A]re there more telephones or animals in the life and works of Hélène Cixous?' in his *H.C. for Life, That Is to Say* …, trans. Laurent Milesi and Stefan Herbrechter (Stanford: Stanford University Press, 2006).
149 Cixous, 'Cat's Arrival', 32.
150 Ibid.
151 Ibid. The persistence of doors and their surprising traversal inevitably suggests – and revises – Kafka's well-known formulation of the condition of the law. Unlike the 'man from the country', the cat passes the door and thus retouches the law. See Kafka, *The Trial*, Johnny Golding drew my attention to the air of Kafka marked by Cixous's door.
152 Numerous writers have commented on the glimmer of recognition offered by 'Bobby the Dog' to the dehumanized prisoners in the camps only for Lévinas to erase an ethical maxim from his barks since the animal is 'too stupid, *trop bête*' to be able to universalize this relation. See, for example, John Llewelyn qtd. Wolfe, *Zoontologies* 17.
153 Cixous, 'Cat's Arrival', 22.
154 Ibid.

155 Ibid.
156 Cary Wolfe, 'Exposures' in Stanley Cavell, Cora Diamond, John McDowell, Ian Hacking and Cary Wolfe, eds. *Philosophy & Animal Life* (New York: Columbia University Press, 2008) 8.
157 See Turner, 'Telefoam'.
158 Cixous, 'Cat's Arrival', 33, *Messie*, 79.
159 Cixous, 'Cat's Arrival', 33.
160 Derrida, *Animal*, 50.
161 Cixous, 'Cat's Arrival', 22.

Chapter 6

1 Jacques Derrida, 'Tympan', *Margins of Philosophy* [1972] trans. Alan Bass (Chicago: Chicago University Press, 1982) x, n.1.
2 Derrida, 'Tympan', x.
3 See, for example, Jacques Derrida, *Of Grammatology*, trans. Gayatri Chakravorty Spivak (Baltimore: Johns Hopkins University Press, 1997) 244; Jacques Derrida, *The Animal That Therefore I Am*, trans. David Wills, ed. Marie-Louise Mallet (New York: Fordham University Press, 2008) 24.
4 The French word makes use of the 'a', as with *différance*. Derrida, 'Tympan', xi. Usefully glossing the inference of Derrida's critique of this central Hegelian movement, Bass notes, 'there is nothing from which the *Aufhebung* cannot profit'. See '*Différance*' in *Margins of Philosophy*, 20, n. 23.
5 Derrida, 'Tympan', xxv.
6 Derrida, *Animal*, 29
7 Derrida, 'Tympan', xxv, emphasis original. The stamp (Fr. *timbre*) on a postcard echoes the vibrating *timbre* of 'Tympan'. See David Wills, *Matchbook: Essays in Deconstruction* (Stanford: Stanford University Press, 2005) 54.
8 Lynn Turner, 'Insect Asides' in Lynn Turner, ed. *The Animal Question in Deconstruction* (Edinburgh: Edinburgh University Press, 2013) 65.
9 Derrida, 'Tympan', xxv.
10 While 'Tympan' attracts relatively modest attention within its detailed pages, John Mowitt's theorization of the 'percussive subject' could be fruitfully read alongside the concerns of this chapter. See John Mowitt, *Percussion: Drumming, Beating, Striking* (Durham: Duke University Press, 2002) 23.
11 *Dancer in the Dark* (2000), [Film] Dir. Lars von Trier, Denmark, Sweden, Netherlands, United States, France, Germany, Italy, Finland, UK, Norway,

Iceland, Argentina, Taiwan, Belgium: Zentropa Entertainments. The trailer is accessible here: https://www.imdb.com/title/tt0168629/. Accessed 20 May 2020.

12 Cary Wolfe, 'When You Can't Believe Your Eyes (or Voice): *Dancer in the Dark*' in *What Is Posthumanism?* (Minneapolis: Minnesota University Press, 2010) 185.

13 See Michel Leiris, [1948] *Scratches: Rules of the Game Volume 1*, trans. Lydia Davis (Baltimore: Johns Hopkins University Press, 1991).

14 Jacques Derrida, *Glas* [1974] trans. John P. Leavey Jr. and Richard Rand (Lincoln: Nebraska University Press, 1986).

15 Derrida, 'Tympan', xii.

16 Ibid., xvii.

17 Ibid., xxiii.

18 Jacques Derrida, 'Typewriter Ribbon: Limited Ink (2)' in *Without Alibi*, trans. Peggy Kamuf (Stanford: Stanford University Press) 73.

19 Veit Erlmann, 'Descartes's Resonant Subject' in *Differences: A Journal of Feminist Cultural Studies*, 22:2–3 (2011): 28

20 Erlmann, 'Descartes's Resonant Subject', 28

21 Derrida only names 'ambush' on xv. See also xxvi.

22 Derrida, 'Tympan', xv.

23 For Lacan's explicit naming of the relation of phallus to logos, see his 'The Signification of the Phallus' in *Ecrits* [1966] trans. Bruce Fink (New York: W. W. Norton, 2006) 685–96.

24 Derrida, 'Tympan', xxvii.

25 Ibid., xxiii.

26 *Dogville* (2003), [Film] Dir. Lars von Trier, Denmark, Sweden, France, Germany, Italy, Finland, UK, Norway: Zentropa Entertainments. *Antichrist* (2009), [Film] Dir. Lars von Trier, Denmark, Sweden, France, Germany, Italy, Poland: Zentropa Entertainments. *Melancholia* (2011), [Film] Dir. Lars von Trier, Denmark, Sweden, France, Germany: Zentropa Entertainments.

27 She wrote that 'it was extremely clear to me when I walked into the actresses profession that my humiliation and role as a lesser sexually harassed being was the norm and set in stone with the director and a staff of dozens who enabled it and encouraged it'. 17 October 2017, @bjork. https://www.facebook.com/bjork/posts/10155782628166460

28 Wolfe's earlier version appeared in *Electronic Book Review* in 2001: http://www.electronicbookreview.com/thread/musicsoundnoise/operatic

29 Jacques Derrida, 'And Say the Animal Responded' in Cary Wolfe, ed. *Zoontologies: The Question of the Animal* (Minneapolis: Minnesota University

Press, 2003). As discussed in Chapter 3 of this volume, Derrida's first major critical essay on Lacan was 'Le Facteur de la Verité' [1975] in *The Post Card: From Socrates to Freud and Beyond*, trans. Alan Bass (Chicago: Chicago University Press, 1987) 411–98.

30 Cary Wolfe, 'Cinders After Biopolitics' in Jacques Derrida, ed. *Cinders*, [1987] trans. Ned Lukacher (Minneapolis: Minnesota University Press, 2014) vii–xxx.

31 It is striking that a number of recent books on Von Trier name 'Woman' as their topic by means of a dedicated use of Lacan and Žižek, yet do not cite Wolfe's perhaps 'impure' engagement. See, for example, Ahmed Elbeshlawy, *Woman in Lars Von Trier's Cinema, 1996–2014* (Basingstoke: Palgrave Macmillan, 2018) or Rex Butler and David Denny, eds. *Lars Von Trier's Women* (London: Bloomsbury, 2017).

32 Judith Butler, *Bodies that Matter: On the Discursive Limits of Sex* (New York & London: Routledge, 1993).

33 Wolfe, 'When You Can't Believe', 205

34 See her subsequent co-authored volume, Judith Butler, Ernesto Laclau and Slavoj Žižek, *Contingency, Hegemony and Universality* (London: Verso, 2000).

35 Kaja Silverman, *The Acoustic Mirror: The Female Voice in Cinema and Psychoanalysis* (Indianapolis: Indian University Press, 1987).

36 Wolfe, 'When You Can't Believe', 185. I refer the reader back to the brief discussion of Derrida's devastating critique of exactly this move in 'Le Facteur de la Verité', discussed in Chapter 4.

37 See Jacques Derrida, '… That Dangerous Supplement …' in *Of Grammatology*, trans. Gayatri Chakravorty Spivak (Baltimore: Johns Hopkins University Press, 1976) 141–64. Slavoj Žižek, 'Self-Interview' in *The Metastases of Enjoyment* (London: Verso, 1994). This is a 'Self-Interview' in which, yes, he both asks and answers his own questions.

38 Jacques Derrida, 'For the Love of Lacan' in *Resistances of Psychoanalysis*, trans. Peggy Kamuf et al. (Stanford: Stanford University Press, 1996) 55.

39 Wolfe, 'When You Can't Believe', 185.

40 Jacques Derrida, *The Animal That Therefore I Am*, trans. David Wills (New York: Fordham University Press, 2008) 36.

41 Jacques Derrida, '"Eating Well" or the Calculation of the Subject', [1988] Peter Connor and Avital Ronell (trans.) *Points … Interviews 1974–1994*, Elizabeth Weber ed. (Stanford: Stanford University Press, 1995) 280, emphasis added. See also Derrida, *Animal* 31–2.

42 Derrida, 'Facteur de la Verité', 474, n.51.

43 Derrida, 'Voice II', 168.

44 Žižek, qtd. in Wolfe 'When You Can't Believe', 199.
45 For the same, albeit more influential, manoeuvre, see Mladen Dolar, *A Voice and Nothing More* (Cambridge: MIT Press, 2006).
46 Derrida 'Dialanguages', trans. Peggy Kamuf in Elizabeth Weber, ed. *Points ... Interviews, 1974–1994* (Stanford: Stanford University Press, 1995) 140.
47 This is part of a pincer movement with Dolar, who accuses Derrida of the same thing, making the same error of 'depriv[ing] voice of its ineradicable ambiguity'. See Dolar, *Voice*, 42.
48 Žižek, qtd. in Wolfe, 'When You Can't Believe', 199, Žižek, 'Self-Interview', 195–196, emphasis added.
49 Derrida, 'Voice II', 163, emphasis original.
50 Ibid.
51 Ibid.
52 We might note that while Selma refuses Jeff's ever-ready attentions as anything beyond friendship, she otherwise articulates herself very much in relation to her imaginary father as the ruse for her son.
53 Sarah Dillon and Sarah Jackson's comments on the first draft of this chapter were particularly useful for this section.
54 Wolfe only mentions the first of these movie scenes, when Cathy's translation of the film for Selma is purely spoken, 187. The second scene emphasizes that the soundtrack that Selma hears, instead of seeing the screen image, is thus also *felt*.
55 Wolfe, 'When you can't believe', 194.
56 Arguably this is not the only reference to *Cabaret* in *Dancer*: Selma's muted transformation of the sound of the oncoming train (prior to her 'I've seen all there is to see' number) contrasts to Sally Bowles's (Liza Minelli) decadently wanton scream in the railway tunnel in front of the innocent newcomer Brian Roberts (Michael York) as a train thunders above their heads.
57 Cornelia Vismann, '"*Rejouer les Crimes*" Theatre vs. Video' in *Cardozo Law Review*, 11:2 (1999): 167.
58 See Vismann, '*Rejouer*', 168. The case in question was the notorious one committed by Corporal Lortie, who ran amok in a government building in Quebec in the 1980s, later stating that the government had his father's face (Vismann, 163).
59 Derrida, 'Voice II', 163.
60 Indeed Drew Daniel writes, 'As everyone knows, you cannot close your ears.' He goes on to cite Lacan: 'In the field of the unconscious, the ears are the only

orifice that cannot be closed.' See Drew Daniel, 'Queer Sound' in *Audio Culture: Readings in Modern Music*, Revised Edition, Christoph Cox and Daniel Warner, eds. (London: Bloomsbury, 2017) 63.
61 Derrida, 'Tympan', xvii. Somewhat allusively he also implies it is to take shelter instead in the uncanny uterus – 'the most familiar of dwellings'.
62 Derrida, 'Tympan', xvii, n.9.
63 Ibid., emphasis added.
64 Vismann, '*Rejouer*', 171.
65 Ibid., 175.
66 Derrida, 'Tympan', xiii.
67 Ibid., xiii, n.5
68 Wolfe argues that *Dancer*'s own message departs from the narrative of Selma's fate: however, in practice he says very little about the scene of her execution other than that she with 'radical passivity' chooses to 'do *nothing*' (193). Moreover, Selma's collapse prior to her execution rather dulls the force that Wolfe attributes to her 'NO!' – as does the guilt regarding Gene's eyesight that motivates her self-sacrifice.
69 Jacques Derrida, 'Eating Well', trans. Peter Connor and Avital Ronell in Elizabeth Weber, ed. *Points ... Interviews, 1974–1994* (Stanford: Stanford University Press, 1995) 278.
70 Jacques Derrida, *Glas* [1974], trans. John P. Leavey, Jr. and Richard Rand (University of Nebraska Press, 1986). Michael Naas, 'Derrida Floruit' in *Derrida Today* 9:1 (2016): 13.
71 Jacques Derrida, *The Death Penalty V. I*, trans. Peggy Kamuf (Chicago: Chicago University Press, 2014), 33, 54, 55, 71, 73, 78, 170, 175.
72 Ibid., 42.
73 Ibid., 145.
74 'Life death' is discussed in Chapter 4 of this volume. See Jacques Derrida, 'To Speculate – on "Freud"' in *The Postcard: From Socrates to Freud & Beyond*, trans. Alan Bass (Chicago: Chicago University Press, 1987) 285. Derrida's resistance to the death penalty by means of his beating heart is discussed in the last chapter of this book – the chapter *in lieu* of conclusion.
75 See Wolfe, 'When You Can't Believe', 193. Wolfe, following Žižek, following Lacan, affirms the 'act as feminine' and as suicidal.
76 See Derrida, *Death Penalty I*, 15 (with regard to Rousseau).
77 I think it is Bodoni bold, as used here. NB. All of Selma's songs were written by Björk.

78 Derrida, 'Tympan', xxv.
79 Derrida, *Animal*, 135, emphasis original.
80 Derrida, 'Dialanguages', 143.

Chapter 7

1 *White God* (2014), [Film] Dir. Kornél Mundruczó, Hungary: Proton Cinema.
2 Film trailer accessible here: http://www.imdb.com/title/tt2844798/?ref_=nm_flmg_act_1
3 Swedish House Mafia, 'Save the World', 2011. https://www.youtube.com/watch?v=BXpdmKELE1k. Accessed 1 August 2017. This video was brought to my attention by Carla Freccero's presentation 'What's in a Name?' at the *Queer Animal* Symposium, King's College, London, June 2012.
4 Derrida, 'Force of Law', trans. Gil Anidjar, in Gil Anidjar, ed. *Acts of Religion* (New York and London: Routledge, 2002) 247.
5 Jacques Derrida, 'Force of Law: The "Mystical Foundation of Authority"', trans. Gil Anidjar, in Gil Anidjar, ed. *Acts of Religion* (New York and London: Routledge, 2002) 252–3. This essay was first published simultaneously in French and English, trans. Mary Quaintance in *Cardozo Law Review*, 11:219 (1990): 920–1039. In 1994 Derrida published a revised French edition and this 'complete' version is the one re-translated for *Acts of Religion*, cited here.
6 Mary Beth Haralovich qtd. in Annette Kuhn, [1982] *Women's Pictures: Feminism and Cinema*, 2nd Edition (London and New York: Verso, 1994) 34. This structural gendering is hardly limited to cinema; see, for example, Susan McClary, *Feminine Endings: Music, Gender and Sexuality* (Minneapolis: Minnesota University Press, 1991).
7 Kuhn, *Women's Pictures*, 34. In a brilliant rejoinder to this entire problem, the very last episode of Hulu television series *Buffy the Vampire Slayer* (after a seven season run) destroyed the conditions of possibility of its own legend while maintaining both the life and sexual independence of its eponymous heroine – now one among many such slayers rather than the exception.
8 Dylan Hallinstad O'Brien, Review of *White God* in *Feral Feminisms*, 6 (2016) 119–22. *Lassie Come Home* (dir. Fred. M. Wilcox, US, 1943) revolved around a young boy and his devoted pet dog; *The Incredible Journey* (dir. Fletcher Markle, US, 1963) is the journey of three pets finding their way home. Erica Fudge, *Pets* (Stocksfield: Ashgate, 2008).

9 Katarina Gregersdotter, Johan Hoglund, Nicklas Hallen, eds. 'Introduction', in *Animal Horror Cinema: Genre, History and Criticism* (Basingstoke: Palgrave Macmillan, 2015) 9.
10 That some refer to this revolutionary uprising as a 'psychotic outbreak' seems wilfully misleading. See, for example, Peter Bradshaw's review in *The Guardian*, 26 February 2015. https://www.theguardian.com/film/2015/feb/26/white-god-review-kornel-mundruczo
11 *Spartacus*, dir. Stanley Kubrick, US, 1960. *Rise of the Planet of the Apes*, dir. Rupert Wyatt, US, 2011.
12 Dinesh Wadiwel, *The War Against Animals* (Leiden: Brill, 2015) 269.
13 Unlike the ape leader, Caesar, as he justifies breaking his own law against the killing of other apes by saying to his foe, 'You are not ape'; see Johan Hogland, 'Simian Horror in *Rise* and *Dawn of the Planet of the Apes*' in *Animal Horror Cinema*.
14 Talionic justice is abbreviated by the biblical phrase 'an eye for an eye'. See Exodus 21.24–5; Leviticus 24.20; Deuteronomy 19.21
15 See Jacques Derrida, '*Che cos'è la poesia?*' in Elizabeth Weber, ed. *Points ... Interviews 1974–1994* (Stanford: Stanford University Press) 288–99.
16 Derrida, '*Che cos'è la poesia?*' 295.
17 Derrida, *The Beast and the Sovereign, V.I*, trans. Geoffrey Bennington (Chicago: Chicago University Press, 2009) 17–18.
18 Colin Dayan, *The Law Is a White Dog: How Legal Rituals Make and Unmake Persons* (Princeton: Princeton University Press, 2011).
19 Dayan, 39, 252. See also Bénédicte Boisseron, 'Afro-dog' in *Transition*, 118 (2015): 15–31. Boisseron's discussion of 'cyno-racial kinship' invokes both Dayan's work and Fuller's film.
20 *White Dog* (Dir. Sam Fuller, 1982, US: Paramount Pictures). Trailer available here: https://www.imdb.com/title/tt0084899/
21 Sheila Roberts, '*White God* Interview: Director Kornel Mundruczo', 27 March 2015. http://collider.com/white-god-kornel-mundruczo-interview/. Accessed 20 May 2020.
22 Jacques Derrida, 'Force and Signification' in *Writing and Difference*, trans. Alan Bass (London: Routledge, 2001) 5, italics original.
23 Derrida, 'Force of Law', 247.
24 John Greco discusses *White Dog* in his blog *Twenty Four Frames*. https://twentyfourframes.wordpress.com/2010/03/06/white-dog-1982-sam-fuller/
25 Romain Gary, *Chien Blanc* (Paris: Éditions Gallimard, 1970).
26 The dog was played by five other dogs (credited as Hans, Folsom, Son, Buster and Duke).

27 Susanne Schwertfeger also suggests the potential correlation between Carruthers and the dog's first trainer; see her 'Re-education as Exorcism: How a White Dog Challenges the Strategies for Dealing with Racism' in Gregersdotter, Hoglund, Hallen, eds. *Animal Horror Cinema*, 138.
28 A great deal of care evidently went into training this dog to 'play dead', with his head softly hitting the ground.
29 Jacques Derrida, 'Signature Event Context', in *Margins of Philosophy*, trans. Alan Bass (Chicago: University of Chicago Press, 1982) 329, emphasis original.
30 Jacques Lacan qtd. in Derrida, *Beast* 1, 97. See Jacques Lacan, 'A Theoretical Introduction to the Functions of Psychoanalysis in Criminology' in *Ecrits*, trans. Bruce Fink (New York: W. W. Norton, 2005) 137.
31 This transubstantiation of animal into human is reminiscent of Freud's interpretation of animal phobias: 'In every case it was the same: the fear was basically of the father, where the children under examination were boys, and had merely been displaced on to the animal.' Freud, 'Totem & Taboo', 128.
32 Derrida, *Beast* 1, 110.
33 Ibid., 107.
34 Jacques Derrida, *The Death Penalty V.I*, trans. Peggy Kamuf (Chicago: Chicago University Press, 2014) 141. These seminars took place from 1999 until 2001.
35 Derrida, *Death Penalty I*, 141.
36 Much publicity and discussion around the film noted that these were real dogs, not CGI ones, running through Budapest, that they were trained with care and adopted by the end of the film's production. See Katherine Tarpinian's interview with the dog trainer Teresa Ann Miller, 'How 200 Dogs Were Trained to Act in "WHITE GOD,"' 15 April, 2015. https://creators.vice.com/en_uk/article/mgp3j8/how-200-dogs-were-trained-to-act-in-white-god
37 Gil Anidjar, 'A Note on "Force of Law"' in Jacques Derrida, ed. *Acts of Religion* (New York and London: Routledge, 2002) 228.
38 Lacan, 'The Mirror Stage as Formative of the *I* Function as Revealed in Psychoanalytic Experience'. [1949] in *Ecrits*, 75–81.
39 See Vicki Hearne, *Adam's Task: Calling Animals by Name* (New York: Sky Horse Publishing, 2007). For a discussion of horse whisperer Monty Roberts's practice of 'starting' rather than 'breaking' horses in light of observing wild mares discipline errant young horses, see Paul Patton, 'Language, Power and the Training of Horses', in Cary Wolfe, ed. *Zoontologies: The Question of the Animal* (Minneapolis: Minnesota University Press, 2003).

40 *The Texas Chainsaw Massacre*, dir. Tobe Hooper, US, 1974. In the film's cannibal family, 'Grandpa' is described as a former slaughterhouse worker, presumably at the plant seen *en route* to the old family homestead.
41 Nicole Shukin, *Animal Capital: Rendering Life in Biopolitical Times* (Minnesota University Press, 2009) 87. At the time of writing, a petition was in circulation against a new level of 'high speed' slaughter and the extreme levels of loss of care in service to the machine and the demands of capital. See Scott David, 'America's Horrifying New Plan for Animals: High-Speed Slaughterhouses' *The Guardian*, 6 March 2018. https://www.theguardian.com/commentisfree/2018/mar/06/ive-seen-the-hidden-horrors-of-high-speed-slaughterhouses.
42 Noëlie Vialles, *Animal to Edible*, trans. J. A. Underwood (Cambridge: Cambridge University Press, 1994) 53–4.
43 Shukin, *Animal Capital*, 91.
44 Ibid., 97–8. Nicole Shukin and Sarah O'Brien, 'Being Struck: On the Force of Slaughter and Cinematic Affect' in Michael Lawrence and Laura McMahon, eds. *Animal Life and the Moving Image* (Basingstoke: BFI Publishing, 2015) 187–202. In an interview, Mundruczó affirmatively compares *White God* to *Strike*, taking the dogs for the workers: ND/NF Interview: Kornél Mundruczó by Yonca Talu, 20 March 2015. https://www.filmcomment.com/blog/ndnf-interview-kornel-mundruczo-white-god/
45 The nadir of this knock-on effect perhaps is in the Maroon 5 pop video, *Like Animals,* in which the self-conscious equation of montage and butchery is seemingly designed to attract young women viewers even as the lead singer is also, for the narrative purpose of the song, both a butcher and a stalker. The blow here instigates submission rather than revolution.
46 As Shukin and O'Brien acknowledge, in light of Jonathan Burt's work on 'pro-animal' films, this logic is hardly confined to Eisenstein.
47 The interior slaughterhouse scenes were filmed in an abattoir exterior to Budapest, the courtyard used for the film is that of the city's National Heritage Protection Centre (correspondence with Eszter Timár, Budapest, January 2018).
48 Vialles notes the loss of skills and loss of confrontation with and acknowledgement of individual animals in *Animal to Edible*, 52.
49 Vialles, *Animal to Edible*, 46.
50 See Charlie LeDuff 'At a Slaughterhouse, some things never die' in Cary Wolfe, ed. *Zoontologies: The Question of the Animal* (Chicago: Chicago University Press, 2003) 183–98 and Timothy Pachirat, *Every Twelve Seconds: Industrialised Slaughter and the Politics of Sight* (New Haven: Yale University Press, 2013).

51 Deuteronomy 14:14, qtd. in Vialles 73.
52 Vialles, 33.
53 The environmental considerations of such a vast and polluting production of waste are discussed here: http://www.huffingtonpost.co.uk/sue-cross/horse-meat-slaugtherhorse-veganism_b_2684502.html
54 Vialles, *Animal to Edible*, 66.
55 Derrida, *Death Penalty I*, 96.
56 Ibid. Italics original.
57 Tobias Menely, 'Red' in Jeffrey Jerome Cohen, ed. *Prismatic Ecology* (Minneapolis: Minnesota University Press, 2014) 24.
58 Alfred Lord Tennyson, *In Memoriam*, 1849.
59 Claire Jean Kim discusses the trainer's abusive trick of maintaining sightlines between dog and owner so that they will continue fighting *for* their owner. See her 'Michael Vick, Dogfighting and the Parable of Black Recalcitrance' in *Dangerous Crossings: Race, Species and Nature in a Multicultural Age* (Cambridge: Cambridge University Press, 2015) 253–82.
60 See Kim, *Dangerous Crossings*.
61 See Gil Anidjar, '*Le Cru*: Derrida's Blood' in theory@buffalo 2015, 8–22.
62 Vialles, *Animal to Edible*, 92.
63 Dawne McCance fleshes out Heidegger's faith that the animal cannot die but merely 'perishes' as a consequence of the animal's inability to apprehend the world as such by means of language. See her 'Death' in Lynn Turner, Undine Sellbach and Ron Broglio, eds. *The Edinburgh Companion to Animal Studies* (Edinburgh: Edinburgh University Press, 2018) 122.
64 Kelly Oliver, 'See Topsy "Ride the Lightning": The Scopic Machinery of Death' in *The Southern Journal of Philosophy*, 50 Spindel Supplement (2012): 76.
65 Or even 'immigrants'. The political situation in Budapest under the ever-more right-wing premiership of Viktor Orban is directing increasing hostility towards refugees. See Joe Wallen, 'Hungary Is the Worst:' Refugees Become Punch-bag Under PM Viktor Orban' *The Independent*, 13 July 2018. https://www.independent.co.uk/news/world/europe/hungary-refugees-immigration-viktor-orban-racism-border-fence-a8446046.html
Orban has also explicitly targeted feminists, banning the teaching of 'Gender Studies' in Hungarian universities. See Maya Oppenheim, 'Hungarian Prime Minister Viktor Orban Bans Gender Studies Programmes' *The Independent*, 24 October 2018. https://www.independent.co.uk/news/world/europe/hungary-bans-gender-studies-programmes-viktor-orban-central-european-university-budapest-a8599796.html

66 Derrida, *Death Penalty I*, 23.
67 Ibid., 231.
68 Oliver, 'See Topsy "Ride the Lightning"', 77.
69 Shukin, *Animal Capital*, 154. As she notes, films of executions were common subjects in early cinema, whether of humans or animals.
70 Shukin, 160.
71 Ibid.
72 This is not simply about abolitionism's partial form that looks only to mitigate 'cruel and unusual punishment' but about the use of drugs to make it appear that no pain has been caused. For example, Oliver, 'See Topsy "Ride the Lightning"'.
73 Derrida, *Death Penalty I*, 276.
74 Elizabeth Rottenberg, 'Deconstructing Death: Derrida and the Scene of Execution' in *New Formations*, 89/90 (2016) 34.
75 Derrida, *Death Penalty I*, 116.
76 Derrida cites Rousseau in this regard; see *Death Penalty I*, 15.
77 Ibid., 116.
78 Ibid., 2, italics original. Derrida is thus in explicit critical dialogue with the Michel Foucault of *Discipline and Punish*; see *Death Penalty I*, 42.
79 Ibid., 2–3.
80 Ibid., 23, emphasis added.
81 Ibid., 42. Oliver expands the theatrical aspects of Derrida's critical implication of 'speculative philosophy' but leaves the question of idealism and the sense of speculation as a gamble out of her analysis; see her 'See Topsy "Ride the Lightning"', 75–6.
82 Derrida, *Death Penalty I*, 258.
83 Ibid.
84 Ibid., 256.
85 Ibid., 257.
86 The one exception is the spoiler-savvy hint to 'a stirring moment of submission' in J. Larsen's 4* review. http://www.larsenonfilm.com/white-god
87 Derrida, *Death Penalty I*, 49.
88 Ibid., 220.
89 Lili only says, 'I love you too' to Hagen much earlier in the film, when she is playing her trumpet to them both in the bathroom of her father's apartment. Then there is no suggestion of disequivalence or a question of the status of their relation.
90 Derrida, '*Che cos'è la poesia?*', 291.
91 Derrida, *Death Penalty I*, 257.

92 Derrida, '*Che cos'è la poesia?*', 297.
93 See Timothy Clark, 'By Heart: A Reading of Derrida's "*Che cos'è la poesia?*" Through Keats and Celan' in *Oxford Literary Review*, 15:1 (2012): 43–78. Clark also remarks on Derrida's radical revision of how we might understand the hedgehog as that entity both turned in and turned out – heteropoetic perhaps – rather than able to stand alone to 'figure the Ideal of the Romantic text as unconditioned', as it did for Schlegel (46).
94 Derrida, '*Che cos'è la poesia?*' 289.
95 Ibid., 299. Nicholas Royle remarks that this 'demon of the heart' is unattributed, thus he takes it as Derrida's name for the poem itself. See Nicholas Royle, *Jacques Derrida* (New York and London: Routledge, 2003) 137.
96 Derrida, 'Force of Law', 270.
97 Derrida, *Beast* I, 108.

Index

abasement 35-9
abattoir 139, 141, 149-50, 156, 204 n.47
ability 24-5, 49, 53, 55, 80, 86, 93, 96-7, 101, 109, 147, 170 n.76
abjection 9, 11, 23, 28-31, 40, 42, 45, 106
abolitionism 3, 206 n.72
Abraham, Nicolas and Maria Torok 16, 18-19, 25
  Absent Presence (Chalayan) 55, 75-80
Adams, Carol 47, 94
agriculture 148, 154
allegory 7, 87, 91, 110, 112, 140, 147, 148
ambivalence 10, 23, 24, 46, 66, 73, 74
animal horror film 151
animal industrial complex 141, 149
Animal Lessons (Oliver) 28, 30
animal phobias 10, 103, 203 n.31
'animal question' 1, 11, 28, 29, 40, 51, 71, 81, 91, 93, 84, 101, 110, 121, 140, 142, 143, 154, 165 n.4
animal studies 4, 28, 30, 81, 90, 102, 148
Animal That Therefore I Am, The (Derrida) 1-4, 9, 24, 80, 81, 83-5, 94, 95, 106, 121, 146
'animal, the' 2, 9, 26, 28, 29, 40, 43, 93-5, 97, 171 n.9, 194 n.133
animality 11, 15, 24, 29, 45, 46, 50, 109, 113, 123, 145, 155
animot 81, 85
anopthalmia 45, 47-9
anthropocentrism 112
anthropomorphism 94, 111, 147, 153
anti-Semitism 17
apostrophe 5, 161, 163
audience 13, 20-2, 24, 34, 45, 61, 62, 70, 108, 127, 130, 131, 134, 135, 142, 146, 149, 159, 163
auto-affection 56, 76, 96
autobiography 54, 57, 58, 60, 86, 109-10
autopsy 25, 152

beast 7, 35, 39, 40, 46-8, 50, 52, 87, 94, 112, 140, 141, 144, 151, 153, 157, 158, 159
Beast and the Sovereign, The (Derrida) 25, 144, 152, 161
Beauvoir, Simone de 28, 29
'Belief Itself' (Irigaray) 64-6, 68, 69, 70
bet 57, 59, 71, 80, 151-2, 156, 163
biology 83, 95, 100, 129
bird song 160
bite/biting 11, 27, 34, 46, 49, 152, 173 n.29
Björk 118, 120-1, 122, 124, 125, 129, 135, 200 n.77
'Blind Spot in an Old Dream of Symmetry, The' 50, 63, 64, 176 n.84
blindness 45, 49, 50, 120, 176 n.84
blood 3, 10, 27, 44, 49-52, 63, 65, 71, 104, 132, 145-59
breast, the 17, 19, 44, 49, 53
brotherhood (and brothers) 8, 10, 17, 73, 103, 104, 109
Butler, Judith 82, 121, 122

cage 33, 44, 49, 71, 143, 153
calculable, the 16, 138, 157
calculation 63, 80, 140, 157, 158
cannibalism 8, 10, 11, 18, 40, 41, 46, 48, 74, 103
Carby, Hazel 38
carnivory (the carnivorous) 17, 18, 24, 46, 144, 176 n.78
carno-phallogocentrism 17, 24, 46, 138, 154, 157
castration 4, 49, 64, 65, 134, 176 n.84
cats 12, 23, 30, 86, 110, 114, 115
'Cat's Arrival, The' (Cixous) 81, 86, 109, 110-15
Chalayan, Hussein 55, 75-80
'Cinders after Biopolitics' (Wolfe) 121
civilization 32, 34-6, 39, 44, 140
'Civilization and Its Discontents' (Freud) 30, 31, 35

Cixous, Hélène 81, 83, 85–6, 93, 109, 110–15
   'Cat's Arrival, The' (Cixous) 81, 86, 109, 110–15
   La (Cixous) 85, 111
   La Jeune Née (Cixous) 111
   Messie (Cixous) 110, 188 n.27
   'From My Menagerie to Philosophy' (Cixous) 85, 188 n.27
   'Writing Blind' (Cixous) 81
classic cinema 138, 153
climate crisis 104
closure 26, 89, 128, 138, 160
clothing 20, 77–9, 111–13, 142, 149
companion species 33, 84, 86, 102, 183 n.3, 193 n.107
compost 84, 105
confession 82, 126
consanguinity 8, 13
   Conceiving Ada (Hershman Leeson) 54–5, 61–3, 69–74
   court 25, 32, 126–32, 156
Companion Species Manifesto (Haraway) 86, 98–101, 102
Creed, Barbara 30, 31
crime 41, 128, 130, 145–7
criminal 10, 22, 40, 43, 90, 132, 145
cruelty 33, 36, 51, 134, 145, 150
cure 7–9, 23–6
curiosity 25–6, 43, 50, 52, 85, 99, 106
cut (cinematic) 58, 94, 99, 135, 138, 150, 151
cyborg 83, 84, 100, 186 n.4
'Cyborg Manifesto' (Haraway) 106

'dark continent' 37, 174 n.42
Dancer in the Dark (von Trier) 118, 120, 122, 124–36
Dasein 95, 97, 169 n.66
death drive 60, 61, 71, 80
death penalty 3, 17, 130–3, 144–6, 149, 153–7, 161
   death by electrocution 154
   death by hanging 131, 132
   execution 125, 131–5, 148, 156, 157, 159, 160, 200 n.68, 206 n.69
Derrida, Jacques 7, 11, 15–19, 24–6, 28–30, 43, 49, 51, 56
   Animal That Therefore I Am, The (Derrida) 1–4, 9, 24, 80, 81, 83–5, 94, 95, 106, 121, 146
   Beast and the Sovereign, The (Derrida) 25, 144, 152, 161
   Death Penalty (Derrida) 3, 17, 132–4, 143–7, 150, 154–8, 161
   'Eating Well' (Derrida) 15–18, 40, 103, 154
   '*Fors*' (Derrida) 19, 167 n.38
   life death 56, 60, 67, 72, 74, 80, 131, 132, 200 n.74
   limitrophy 26, 51, 115
   machine-event 82–3, 119, 124
   'Plato's Pharmacy' (Derrida) 79–80
   sexual difference 2, 9, 11, 47, 48, 63, 65, 67, 69, 89
   'To Speculate – on "Freud"' (Derrida) 54, 56–61, 68–9
   trace, the 12, 24–6, 83, 121, 124, 136
   'Tympan' (Derrida) 117–20, 124, 129, 131, 135, 196 n.4
   'Typewriter Ribbon' (Derrida) 82–3, 119
devourment 39, 48, 49
dialectics 5, 9, 43, 56, 63, 68, 113
   master-slave dialectic 43
difference 5, 9–11, 37, 38, 63–5, 99, 111, 117, 121, 122
   animal 2, 11
   racial 36
   sexual 2, 9, 11, 47, 48, 63, 65, 67, 69, 89
disability 4, 48, 50, 177 n.92
disgust 14, 15, 23, 44
dogfighting 144–7, 151, 153, 205 n.59
dogs 7, 8, 14, 15, 21, 26, 33–5, 44–7, 49, 50, 86, 99, 101, 102, 104, 137, 139, 140, 142, 146–8, 151–4, 158–60, 163, 169 n.68, 173 n.33, 176 n.78
Dolar, Mladen 121, 199 n.47
domestic violence 31, 46
domestication 33, 51, 66, 102, 177 n.94
Dunsany, Lord 7, 164 n.2

ear, the 18, 20, 118, 119, 120, 128, 129, 131, 133, 134, 199 n.60
'eating the other' 16, 18, 23, 183 n.78
'Eating Well' (Derrida) 15–18, 40, 103, 154
economy of exchange 11, 27, 32, 33, 37, 38–9, 40, 42–3, 48, 146, 175 n.66
elephants 25, 153–5, 170 n.76, 170 n.78

Enlightenment, the 36, 145
essentialism 69, 132, 184 n.100
ethics 7, 9, 15, 16, 18, 39, 65, 122, 162, 185 n.3, 190 n.72
eugenics 104, 105, 185 n.120
evolution 31, 95, 166 n.16
'Exposures' (Wolfe) 113
Eye/Body (Schneemann) 89, 115

fable 7, 25
Facebook 62, 121, 181 n.56
family, the 10, 31–4, 35, 36, 38, 40, 50, 51, 62, 103–4, 132, 139, 140
father, the 8, 10, 11, 12, 13, 17, 23, 26, 31, 38, 41–2, 49, 73, 100, 128, 139, 144–4, 148, 150, 156, 159, 160, 176 n.82, 203 n.31
fellow 26, 145, 151
feminine, the 15, 28, 29, 31, 32, 41, 42, 43, 44, 48, 50, 54, 64–5, 68, 74, 90, 118, 122–4, 129, 131
feminism 3, 28, 54, 57, 62, 73, 97, 100, 119
fetish 3, 34, 37, 69, 120, 172 n.25
finitude 107, 153, 160
food 11, 13–18, 24, 33, 40, 44, 47, 103, 140, 150
fort/da 53–4, 56–9, 61, 64, 67–8, 80
'Fors' (Derrida) 19, 167 n.38
fraternity 24, 35, 43, 104
Freud, Sigmund 14, 15–18, 30–2, 35–7, 47, 48, 70, 86, 100, 145, 166 n.16
  'Beyond the Pleasure Principle' (Freud) 53, 54, 57–61, 63, 65–9, 70, 71, 76, 117, 167 n.32
  'Civilization and Its Discontents' (Freud) 30, 31, 36, 40–2
  'Femininity' (Freud) 63
  'Taboo of Virginity' (Freud) 3–4
  Totem and Taboo (Freud) 8, 9, 10, 11, 17, 27, 41, 43, 73, 145
Freud, Sigmund and Josef Breuer 8
Fuller, Sam 142, 202 n.19
Fuses (Schneemann) 89, 90

gaze, the 86, 89, 90, 98, 106, 109, 114, 115, 122, 175 n.57
gender 5, 32, 33, 37, 38, 40, 48, 65, 72, 79, 90, 91, 99, 100, 111, 184 n.100, 193 n.127, 201 n.6, 205 n.65

'"Gender" for a Marxist Dictionary' (Haraway) 5, 38, 100, 174 n.47
genocide 193 n.107
Geschlecht 5, 38, 174 n.46
'Gesture in Psychoanalysis' (Irigaray) 54, 63, 65, 67, 68, 69
girl, the 10, 37, 52, 54, 63–9, 74, 100, 183 n.97
Goodeve, Thryza Nichols 99, 188 n.29, 189 n.41
Guyer, Sara 16, 169 n.50

hand, the 86, 95–7, 107, 114
Haralovich, Mary Beth 138
Haraway, Donna 5, 33, 57, 81–4, 85, 93, 167 n.27
  Companion Species Manifesto (Haraway) 86, 98–101, 102
  'Cyborg Manifesto' (Haraway) 106
  '"Gender" for a Marxist Dictionary' (Haraway) 5, 38, 100, 174 n.47
  How Like a Leaf (Haraway) 193 n.114
  Modest_Witness (Haraway) 104, 106, 108
  'Nothing Comes without Its World' (Haraway) 193–4 n.127, 194 n.128
  Primate Visions (Haraway) 186 n.4
  Staying with the Trouble (Haraway) 105, 193 n.107, 193 n.109
  When Species Meet (Haraway) 94, 101, 192 n.91
heart, the 1, 2, 3, 4, 49, 51, 112, 114, 132, 139–41, 146, 151, 157, 158, 160, 161–2, 163
hedgehog 1, 2, 161, 164 n.5, 188 n.27, 207 n.93
Heidegger, Martin 11, 22, 95–8, 101, 107, 113, 169 n.66, 174 n.46, 190 n.68
Hershman-Leeson, Lynn 84, 178 n.10
heterosexuality 37, 61, 102, 138
home, the 16, 21, 22, 23, 33, 35, 46, 60, 61, 62, 67, 70, 110, 139, 140, 150, 173 n.30
homosociality 15
horizontality 15, 20, 31, 48, 65, 115, 146, 160
horror film 31, 34, 44, 49, 140, 148, 151, 172 n.19, 173 n.29
hospitality 9, 24, 51, 85, 110, 112
  infinite hospitality 16, 18

How Like a Leaf (Haraway) 193 n.114
humanism 18, 84, 100, 123, 177 n.92
human, the 2, 9, 28, 44, 83, 93–5, 146
hysteria 8, 68, 167 n.32

identification 16, 18, 40, 48, 62, 141, 147, 148, 153
immigrant (and immigration) 17, 118, 120, 128, 132, 185 n.120, 205 n.65
'Impudence of Uttering, The' (Kristeva) 29–30
incalculable, the 91, 101, 158
incest 10, 11, 41, 42, 102, 103, 104, 175 n.66
incorporation 19, 66, 67
Infinity Kisses I (Schneemann) 86–91, 93, 114–15
Infinity Kisses II (Schneemann) 87, 89, 114–15
Infinity Kisses – The Movie (Schneemann) 87
ingestion 16, 18, 20, 25, 26, 48, 91
Interior Scroll (Schneemann) 86, 188 n.28
introjection 16–19, 66
Irigaray, Luce 53, 63–9, 74, 80, 119, 176 n.84, 181 n.61, 182 n.62 63 64, 184 n.100, 189 n.48
 'Belief Itself' (Irigaray) 64–6, 68, 69, 70
 'Blind Spot in an Old Dream of Symmetry, The' (Irigaray) 50, 63, 64, 176 n.84, 183 n.97
 'Gesture in Psychoanalysis' (Irigaray) 54, 63, 65, 67, 68, 69, 183 n.76, 88 89

jus talionis 140, 144–7, 159
justice 128, 130, 131, 135, 137, 139, 141, 142, 145–7, 161, 202 n.14

kennels 33–5, 44, 46–50
Khanna, Ranjanna 37, 166 n.9
kinship 10, 51, 93, 94, 103, 104, 175 n.58
 kith and kin 93
kiss 81, 82, 86–93, 98, 99, 101, 104, 106, 109, 112, 114, 173 n.29
kithe 92–7, 101, 114
Klein, Melanie 18, 184 n.113
Kristeva, Julia 15, 28–32, 36, 40–5, 48, 171 n.9–11, 175 n.66

'Impudence of Uttering, The' (Kristeva) 29–30
Powers of Horror (Kristeva) 10–11, 28, 30, 41, 42
Kuhn, Annette 138

La (Cixous) 85, 111
La Jeune Née (Cixous) 111
Lacan, Jacques 37, 42, 63, 64, 108, 119–21, 128, 144
lack 3, 18, 19, 23, 37, 41, 100, 121, 122, 132, 134, 141, 181 n.55
'Learning from Temple Grandin' (Wolfe) 49, 177 n.92
legacy 11, 36, 54, 55, 61, 63–9, 83, 102
Levinas, Emmanuel 17, 18, 94, 111, 168 n.43
Levi-Strauss, Claude 32, 37, 42
life death 56, 60, 67, 72, 74, 80, 131, 132, 200 n.74
limitrophy 26, 51, 115
Lipitt, Akira 14

McCance, Dawne 56, 72, 80, 95, 170 n.78, 179 n.14, 185 n.120
McIntosh, Pollyanna 32, 39, 172 n.23, 177 n.90
Mckee, Lucky 31, 173 n.29
maternal, the 23, 28, 29, 33, 42, 45, 49, 51, 73, 74, 134, 145, 176 n.72
Meat Joy (Schneemann) 89
melancholia 7, 18–19, 25
Messie (Cixous) 110, 188 n.27
metaphor 12–13, 18–20, 23, 91, 138, 188 n.37, 189 n.48
metonymy 3, 13, 16, 29, 44, 53, 66, 91, 102, 105, 107, 151, 157, 189 n.48
Miller, Hillis J. 91–2
Modest_Witness (Haraway) 104, 106, 108
modesty 108, 112
montage 47, 148, 204 n.45
mother, the 11, 19, 35, 41, 45, 48, 73
mourning 7–9, 18–19, 22, 24, 91, 133
mouth, the 18, 19, 27, 40, 44–51, 123
Mundruczó, Kornél 137, 142, 204 n.44
murder 28, 39, 41, 42, 51, 60, 73, 126, 144, 159
My Talks with Dean Spanley (Fraser) 7–9, 13, 14, 24–6, 164 n.2

'From My Menagerie to Philosophy'
    (Cixous) 85, 188 n.27

nakedness 89, 106-7, 106-9, 113-14
    naked truth 1, 107, 113, 114
    naked words 1, 2, 108
    nudity 1, 2, 106-7, 113, 114
Nancy, Jean-Luc 16, 17, 84
Nixon, Mignon 72-3, 184 n.113
nose, the 12, 13, 14, 20, 167 n.32
'Nothing Comes without Its World'
    (Haraway) 193-4 n.127, 194 n.128

Oedipus complex 10, 40, 41, 102, 104, 174 n.42
olfaction 14, 152, 167 n.32
Oliver, Kelly 10, 11, 30, 31, 41, 170 n.4, 170 n.76, 171 n.13, 176 n.82, 206 n.81
    Animal Lessons (Oliver) 28-30, 46, 51
    'See Topsy "Ride the Lightning"'
        (Oliver) 153, 154

pardon 133, 159, 160
paternal, the 11, 24, 45, 51, 74, 141, 163
    paternal signifer 42, 47
paternity 15, 23
patriarchy (and patriarchal) 10, 32, 38, 44, 60, 65, 105, 145
patricide 41, 43, 128, 162
performative (and performativity) 82, 83, 92, 94-5, 103, 117, 132, 140, 147, 154, 161, 162
pet, the 8, 12, 23-4, 25, 33, 51, 102-4, 106, 137, 139, 151, 169 n.65
phallogocentrism 17, 80, 120, 154
phallus, the 17, 64, 121-4, 156, 176 n.84, 197 n.23
pleasure principle 54, 57, 59, 60, 63, 68, 76
poematic, the 2, 140, 160
point-of-view shot 152
population 104, 105
'postal principle' 54, 56-61
posthumanism 83, 84
'posthumanist feminine' 118, 122-4, 132, 134
Powers of Horror (Kristeva) 10-11, 28, 30, 41, 42
Powici, Christopher 30, 171 n.14
primal feast 8-10, 15, 24

Primate Visions (Haraway) 186 n.4
privation 49, 53, 93, 94-7, 101, 109, 178 n.5

race 36, 38, 143, 173 n.39
racism 77, 142, 144, 155
rape (and sexual assault) 31, 32, 35, 121, 175 n.57
reaction and response 24, 83, 94, 96, 115, 121, 147
reality principle (RP) 59
rectitude 4, 50, 163
revenge (and vengeance) 31, 48, 49, 146, 159, 160, 162
revolution 37, 140, 148, 149, 158, 159, 161, 202 n.10, 204 n.45
rhythm 2, 54, 118, 125, 129, 130, 131, 133, 134
Rickels, Laurence 24
Rose, Deborah Bird 10
Rubin, Gayle 37-8, 42

sacrifice 16, 17, 24, 43, 50, 51, 133, 134, 155, 200 n.68
science fiction (SF) 54, 55, 57, 106
Schneemann, Carolee 81, 83, 85-93, 114-15
    Eye/Body (Schneemann) 89, 115
    Fuses (Schneemann) 89, 90
    Infinity Kisses I (Schneemann) 86-91, 93, 114-15
    Infinity Kisses II (Schneemann) 87, 89, 114-15
    Infinity Kisses – The Movie
        (Schneemann) 87
    Interior Scroll (Schneemann) 86, 188 n.28
    Meat Joy (Schneemann) 89
    Vesper's Stampede to My Holy Mouth
        (Schneemann) 90
Schneider, Rebecca 90-1
'sex/gender system' 37, 38, 100
sexual difference 2, 9, 11, 32, 47, 48, 63-5, 67, 69, 90, 119, 123, 184 n.100
sexual differences 2, 9, 11, 65, 69, 119, 124
shame 2, 37, 45, 50, 94, 106-8, 110, 112-13, 127
Shell, Marc 102-4, 106
Shukin, Nicole 148-9, 151, 154-5, 206 n.69

sight, sense of 15, 50, 134, 150, 152
  sightline 150, 152, 205 n.59
signature 57, 64, 74, 93, 95, 97, 107, 114, 160
signification 14, 18, 28, 29, 37, 40, 42, 64, 121, 130, 135, 141, 175 n.67
signifier 3, 47, 64, 83, 118, 121, 122, 125, 136, 138, 145
  transcendental signifier 42
slaughter 139, 149, 150, 154, 159, 160, 168 n.44, 204 n.41
slaughterhouse 137, 139, 140, 141, 147, 148, 149, 150, 151, 155, 156, 159, 161, 204 n.40, 204 n.47
smell, sense of 13–15, 20, 21, 26, 31, 36, 152–3
sovereign 25, 39, 40, 46, 133, 141, 144, 151, 153, 157, 159, 160, 170 n.78
sovereignty 7, 82, 101, 133, 144, 153, 154, 157, 160
spacing 96, 97, 98, 124, 134
species 21, 28, 30, 32, 33, 36, 57, 66, 84, 99, 115, 147
species grid 33, 39, 47
'To Speculate – on "Freud"' (Derrida) 54, 56–61, 68–9
speech 18, 19, 91, 123, 189 n.48
stage 132, 155, 158
staggered analogy 97–8
Staying with the Trouble (Haraway) 105, 193 n.107, 193 n.109
'Subject to Sacrifice' (Wolfe) 31, 33, 39, 47
subjectivity 28, 29, 82, 96, 119, 123, 128, 194 n.134
sucker punch 45, 49
suffer, to (and suffering) 49, 136, 147, 153, 155
symbiogenesis 99, 105

technicity 107, 133
technology 77, 113, 155
telephone 91, 133, 159, 195 n.148
telos 53, 55, 100
theatre 130, 135, 156
thinking 85, 95–6, 112–13
totalitarianism 102
Totem and Taboo (Freud) 8, 10, 11, 17, 41, 43, 144
totemism 9–15, 24, 42
trace, the 12, 24–6, 83, 121, 124, 136

transference 37, 61, 65, 72–4, 184 n.113
tympanum 117–20, 129, 131, 132, 135
'Tympan' (Derrida) 117–20, 124, 129, 131, 135, 196 n.4
'Typewriter Ribbon' (Derrida) 82–3, 119

uncanny, the 13, 18, 20, 39, 60, 85, 104, 110, 141, 195 n.141, 200 n.61
unconscious, the 25, 39, 72, 86, 100, 104–5, 123, 199 n.60
urine 12, 13, 26, 167 n.27

verticality 3, 15, 31, 50, 65, 115, 159, 160
Vesper's Stampede to My Holy Mouth (Schneemann) 90
Vialles, Noëlie 148, 150, 152–3
violence 25, 31–2, 45, 46, 51–2, 143, 161
vision 25, 50, 84, 177 n.92
Vismann, Cornelia 82, 127–30, 199 n.58
voice 111, 119, 121–2, 124, 125, 129, 133, 135, 177 n.89, 199 n.47
  voice-over 74, 76–9
vulnerability 2, 4, 79, 101, 136, 147, 160, 161
vulva 127

When Species Meet (Haraway) 94, 101, 192 n.91
'When You Can't Believe Your Eyes' (Wolfe) 118–25, 126, 129
White Dog (Fuller) 142, 202 n.19
White God (Mundruczó) 137–48, 151–4, 156, 158–63
wolf/wolves 27, 39, 40
Wolfe, Cary 198 n.31
  'Cinders after Biopolitics' (Wolfe) 121
  'Exposures' (Wolfe) 113
  'Learning from Temple Grandin' (Wolfe) 49, 177 n.92
  'Subject to Sacrifice' (Wolfe) 31, 33, 39, 47
  'When You Can't Believe Your Eyes' (Wolfe) 118–25, 126, 129, 199 n.54
'woman question' 135, 137–40, 146–51, 154, 155, 158, 160, 204 n.47
Woman, The (Mckee) 27, 31–4, 47, 48, 50–2
'Writing Blind' (Cixous) 81

Žižek, Slavoj 121–4, 200 n.75

Lightning Source UK Ltd.
Milton Keynes UK
UKHW020056230522
403375UK00003B/92